Computational Technologies in Materials Science

Science, Technology, and Management Series

Series Editor:
J. Paulo Davim
*Professor, Department of Mechanical Engineering,
University of Aveiro, Portugal*

This book series focuses on special volumes from conferences, workshops, and symposiums, as well as volumes on topics of current interests in all aspects of science, technology, and management. The series will discuss topics such as mathematics, chemistry, physics, materials science, nanosciences, sustainability science, computational sciences, mechanical engineering, industrial engineering, manufacturing engineering, mechatronics engineering, electrical engineering, systems engineering, biomedical engineering, management sciences, economical science, human resource management, social sciences, engineering education, and so on. The books will present principles, model techniques, methodologies, and applications of science, technology, and management.

Understanding CATIA
A Tutorial Approach
Kaushik Kumar, Chikesh Ranjan, and J. Paulo Davim

Manufacturing and Industrial Engineering
Theoretical and Advanced Technologies
Pakaj Agarwal, Lokesh Bajpai, Chandra Pal Singh, Kapil Gupta, and J. Paulo Davim

Multi-Criteria Decision Modeling
Applicational Techniques and Case Studies
Rahul Sindhwani, Punj Lata Singh, Bhawna Kumar, Varinder Kumar Mittal, and J. Paulo Davim

High-k Materials in Multi-Gate FET Devices
Shubham Tayal, Parveen Singla, and J. Paulo Davim

Advanced Materials and Manufacturing Processes
Amar Patnaik, Malay Kumar, Ernst Kozeschnik, Albano Cavaleiro, J. Paulo Davim, and Vikas Kukshal

Computational Technologies in Materials Science
Shubham Tayal, Parveen Singla, Ashutosh Nandi, & J. Paulo Davim

For more information about this series, please visit: https://www.routledge.com/Science-Technology-and-Management/book-series/CRCSCITECMAN

Computational Technologies in Materials Science

Edited by
Shubham Tayal
Parveen Singla
Ashutosh Nandi
J. Paulo Davim

CRC Press
Taylor & Francis Group
Boca Raton London New York

CRC Press is an imprint of the
Taylor & Francis Group, an **informa** business

First edition published 2022
by CRC Press
6000 Broken Sound Parkway NW, Suite 300, Boca Raton, FL 33487-2742

and by CRC Press
2 Park Square, Milton Park, Abingdon, Oxon, OX14 4RN

© 2022 Taylor & Francis Group, LLC

CRC Press is an imprint of Taylor & Francis Group, LLC

Reasonable efforts have been made to publish reliable data and information, but the author and publisher cannot assume responsibility for the validity of all materials or the consequences of their use. The authors and publishers have attempted to trace the copyright holders of all material reproduced in this publication and apologize to copyright holders if permission to publish in this form has not been obtained. If any copyright material has not been acknowledged please write and let us know so we may rectify in any future reprint.

Except as permitted under U.S. Copyright Law, no part of this book may be reprinted, reproduced, transmitted, or utilized in any form by any electronic, mechanical, or other means, now known or hereafter invented, including photocopying, microfilming, and recording, or in any information storage or retrieval system, without written permission from the publishers.

For permission to photocopy or use material electronically from this work, access www.copyright.com or contact the Copyright Clearance Center, Inc. (CCC), 222 Rosewood Drive, Danvers, MA 01923, 978-750-8400. For works that are not available on CCC please contact mpkbookspermissions@tandf.co.uk

Trademark notice: Product or corporate names may be trademarks or registered trademarks and are used only for identification and explanation without intent to infringe.

Library of Congress Cataloging-in-Publication Data
Names: Tayal, Shubham, editor. | Singla, Parveen, editor. |
Nandi, Ashutosh, editor. | Davim, J. Paulo, editor.
Title: Computational technologies in materials science / edited by Shubham
Tayal, Parveen Singla, Ashutosh Nandi, & J. Paulo Davim.
Description: First edition. | Boca Raton, FL : CRC Press, [2022] |
Series: Science, technology, and management | Includes bibliographical
references and index.
Identifiers: LCCN 2021022140 (print) | LCCN 2021022141 (ebook) |
ISBN 9780367640576 (hardback) | ISBN 9780367640583 (paperback) |
ISBN 9781003121954 (ebook)
Subjects: LCSH: Materials—Data processing.
Classification: LCC TA404.23 .C655 2022 (print) | LCC TA404.23 (ebook) |
DDC 620.1/10285—dc23
LC record available at https://lccn.loc.gov/2021022140
LC ebook record available at https://lccn.loc.gov/2021022141

ISBN: 978-0-367-64057-6 (hbk)
ISBN: 978-0-367-64058-3 (pbk)
ISBN: 978-1-003-12195-4 (ebk)

DOI: 10.1201/9781003121954

Typeset in Times
by codeMantra

Contents

Preface vii
Editors ix
Contributors xi

Chapter 1 Fabrication and Characterization of Materials 1

Laxman Raju Thoutam

Chapter 2 Application to Advanced Materials Simulation 19

Soumyajit Maitra, Souhardya Bera, and Subhasis Roy

Chapter 3 Molecular Dynamics Simulations for Structural Characterization and Property Prediction of Materials 49

S. K. Joshi

Chapter 4 Desirability Approach-Based Optimization of Process Parameters in Turning of Aluminum Matrix Composites 71

P.V. Rajesh and A. Saravanan

Chapter 5 Spark Plasma-Induced Combustion Synthesis, Densification, and Characterization of Nanostructured Magnesium Silicide for Mid Temperature Energy Conversion Energy Harvesting Application 101

P. Vivekanandhan, R. Murugasami, and S. Kumaran

Chapter 6 The Role of Computational Intelligence in Materials Science: An Overview 125

J. Ajayan, Shubham Tayal, Sandip Bhattacharya, and L. M. I Leo Joseph

Chapter 7 Characterization Techniques for Composites using AI and Machine Learning Techniques 139

Dr. Zeeshan Ahmad, Dr. Hasan Zakir Jafri, and Dr. R. Hafeez Basha

Chapter 8	Experimental Evaluation on Tribological Behavior of TiO_2 Reinforced Polyamide Composites Validated by Taguchi and Machine Learning Methods 169
	S. Sathees Kumar, Ch. Nithin Chakravarthy, and R. Muthalagu
Chapter 9	Prediction of Compressive Strength of SCC-Containing Metakaolin and Rice Husk Ash Using Machine Learning Algorithms 193
	Sejal Aggarwal, Garvit Bhargava, and Parveen Sihag
Chapter 10	Predicting Compressive Strength of Concrete Matrix Using Engineered Cementitious Composites: A Comparative Study between ANN and RF Models 207
	Nitisha Sharma, Mohindra Singh Thakur, Viola Vambol, and Sergij Vambol
Chapter 11	Estimation of Marshall Stability of Asphalt Concrete Mix Using Neural Network and M5P Tree 223
	Ankita Upadhya, Mohindra Singh Thakur, Siraj Muhammed Pandhian, and Shubham Tayal

Index 237

Preface

Advanced materials are essential to economic security and human well-being, with applications in industries aimed at addressing challenges in clean energy, national security, and human welfare. However, it can take 20 or more years to move a material after its initial discovery to the market. The computational intelligence involving the use of various computational techniques has accelerated the exploration of materials. The recent years have experienced a burgeoning growth in the development of computational intelligence within the domains of materials science. This book addresses topics related to various computational technologies like conventional microscopy and spectroscopy techniques, AI, and machine learning in materials science. Beginning with the traditional or conventional techniques of material fabrication and characterization, the book also covers the challenges toward developing and using computational intelligence in materials science. This book has been organized into 11 chapters. Chapter 1 gives an overview of the experimental construction and operation principles of different synthesis techniques to make an array of material systems at different length scales (centimeter to nanometer). Chapter 2 offers a brief insight into the various functionals and parameters that are used in the simulation for advanced materials. It explains the density functional theory (DFT) and molecular dynamics (MD) simulations by considering a large range of examples giving deep insights into how the different parameters affect the output results. Chapter 3 features the use of classical MD simulator Large-Scale Atomic/Molecular Massively Parallel Simulator (LAMMPS) along with few other tools like Atomsk, Visual Molecular Dynamics (VMD), and Open Visualization Tool (OVITO) for simulating the structure of materials. A software-assisted optimization that relies on the generation of linear regression equation with the creation of experimental runs known as Response Surface Methodology (RSM) is presented in Chapter 4 to identify the best combination of process parameters and optimized test results. Chapter 5 deals with the synthesis, densification, and characterization of nanostructured magnesium silicide using very basic techniques. Chapter 6 provides a comprehensive overview of the role of computational technologies in materials science. Chapter 7 emphasizes the microstructure characterization of new developing composites and the use of AI and ML techniques for image processing and recognition used for material characterization. Chapter 8 reports the formulation, categorization, and test experimental determination of polyamide6 (PA6) strengthened with titanium oxide (TiO_2) composites using the Taguchi and machine learning methods. Chapter 9 is devoted to highlighting the use of machine learning approaches, viz., Support Vector Machine (SVM), Multilayer Perceptron (MLP), Gaussian Processes Regression (GPR), Random tree, M5P tree, and Random Forest (RF), to develop models to make a reasonable prediction of compressive strength for concrete mix ratios. Finally, Chapters 10 and 11 highlight the use of the various soft computing-based models like artificial neural networks, random forest, and M5P models in predicting the compressive strength of concrete and in evaluating the Marshall stability of asphalt concrete mix. We are grateful to the contributors of each chapter from renowned institutes and industries, and editorial and production teams for their unconditional support to publish this book.

Editors

Dr. Shubham Tayal is working as an Assistant Professor in the Department of Electronics and Communication Engineering at SR University, Warangal, Telangana, India. He has more than 6 years of academic/research experience of teaching at the UG and PG levels. He has received his Ph.D. in Microelectronics & VLSI Design from the National Institute of Technology, Kurukshetra, M.Tech. (VLSI Design) from YMCA University of Science and Technology, Faridabad, and B.Tech. (Electronics and Communication Engineering) from MDU, Rohtak. He has published more than 25 research papers in various international journals and conferences of repute and many papers are under review. He is on the editorial and reviewer panel of many SCI/SCOPUS-indexed international journals and conferences. Currently, he is the editor (or co-editor) for six books from CRC Press (Taylor & Francis Group, USA). He acted as the keynote speaker and delivered professional talks on various forums. He is a member of various professional bodies like IEEE, IRED, etc. He is on the advisory panel of many international conferences. He is a recipient of the Green ThinkerZ International Distinguished Young Researcher Award 2020. His research interests include simulation and modeling of multi-gate semiconductor devices, device-circuit co-design in digital/analog domain, machine learning, and IoT.

Dr. Parveen Singla is a Professor in the Electronics & Communication Engineering Department of Chandigarh Engineering College—Chandigarh Group of Colleges, Landran, Mohali, Punjab. He received his Bachelor of Engineering in Electronics & Communication Engineering with honors from Maharishi Dayanand University, Rohtak; Master's in Technology in Electronics & Communication Engineering with honors from Kurukshetra University, Kurukshetra; and Ph.D. in Communication Systems from IKG Punjab Technical University, Jalandhar, India. He has 17 years of experience in the field of teaching and research. He has published more than 35 papers in various reputed journals, and national and international conferences. He also organized more than 30 technical events for the students in order to enhance their technical skills and received the Best International Technical Event Organiser Award. He is the guest editor of various reputed journals. His interest area includes Drone Technology, Wireless Networks, Smart Antenna, and Soft Computing.

Dr. Ashutosh Nandi has completed his Master's in National Institute of Technology Hamirpur, India, and Ph.D. in Indian Institute of Technology Roorkee, India in the years 2010 and 2015, respectively. At present, Dr. Nandi is working as an Assistant Professor in the Department of Electronics and Communication Engineering, National Institute of Technology Kurukshetra, India. His research area includes low power and low-temperature VLSI design, device circuit co-design in the digital/analog domain, and device modeling of advanced semiconductor devices. To date, Dr. Nandi has guided 2 Ph.D. students and 17 M.Tech. students in the field of VLSI and Semiconductor Devices, has 1 patent, published 31 articles in high-impact-factor journals, and presented more than 25 papers in conferences and proceedings.

J. Paulo Davim is a Full Professor at the University of Aveiro, Portugal. He is also distinguished as an honorary professor in several universities/colleges/institutes in China, India, and Spain. He received his Ph.D. degree in Mechanical Engineering in 1997, M.Sc. degree in Mechanical Engineering (materials and manufacturing processes) in 1991, Mechanical Engineering degree (5 years) in 1986, from the University of Porto (FEUP), the Aggregate title (Full Habilitation) from the University of Coimbra in 2005, and the D.Sc. (Higher Doctorate) from London Metropolitan University in 2013. He is a Senior Chartered Engineer by the Portuguese Institution of Engineers with an MBA and Specialist titles in Engineering and Industrial Management as well as in Metrology. He is also Eur Ing by FEANI-Brussels and Fellow (FIET) of IET-London. He has more than 30 years of teaching and research experience in Manufacturing, Materials, Mechanical, and Industrial Engineering, with special emphasis on Machining & Tribology. He has also interest in Management, Engineering Education, and Higher Education for Sustainability. He has guided large numbers of postdoc, Ph.D., and Master's students as well as has coordinated and participated in several financed research projects. He has received several scientific awards and honors. He has worked as an evaluator of projects for ERC-European Research Council and other international research agencies as well as examiner of Ph.D. thesis for many universities in different countries. He is the Editor in Chief of several international journals, Guest Editor of journals, Editor of books, Editor of book series, and Scientific Advisory for many international journals and conferences. Presently, he is an Editorial Board member of 30 international journals and acts as a reviewer for more than 100 prestigious Web of Science journals. In addition, he has also published as editor (and co-editor) more than 200 books and as author (and co-author) more than 15 books, 100 book chapters, and 500 articles in journals and conferences (more than 280 articles in journals indexed in Web of Science core collection/h-index 59+/11,000+ citations, SCOPUS/h-index 62+/13,500+ citations, Google Scholar/h-index 81+/22,500+ citations). He has been listed in World's Top 2% Scientists by a Stanford University study.

Contributors

Sejal Aggarwal
School of Architecture and Design
Manipal University
Jaipur, India

Zeeshan Ahmad
Department of Mechanical Engineering
Al Falah University
Faridabad, India

J. Ajayan
Department of Electronics
  and Communication Engineering
SR University
Warangal, India

R. Hafeez Basha
Bhasha Research Corporation
Singapore

Souhardya Bera
Department of Chemical Engineering
University of Calcutta
Kolkata, India

Garvit Bhargava
Department of Civil Engineering
National Institute of Technology
  Kurukshetra
Kurukshetra, India

Sandip Bhattacharya
HiSIM Research Center
Hiroshima University
Hiroshima, Japan

Ch. Nithin Chakravarthy
Department of Mechanical Engineering
CMR institute of Technology
Hyderabad, India

Hasan Zakir Jafri
Department of Mechanical
  Engineering
Ashoka Institute of Engineering
  and Technology
Hyderabad, India

L. M. I Leo Joseph
Department of Electronics and
  Communication Engineering
SR University
Warangal, India

S. K. Joshi
Department of Physics
University of Petroleum and Energy
  Studies
Dehradun, India

S. Kumaran
Department of Metallurgical
  and Materials Engineering
National Institute of Technology
Tiruchirapalli, India

Soumyajit Maitra
Department of Chemical Engineering
University of Calcutta
Kolkata, India

R. Murugasami
Department of Metallurgical
  and Materials Engineering
National Institute of Technology
Tiruchirapalli, India

R. Muthalagu
Department of Mechanical Engineering
B.V. Raju Institute of Technology
Narsapur, India

Siraj Muhammed Pandhian
General Studies Department
University College Jubail
Jubail, Saudi Arabia

P.V. Rajesh
Department of Mechanical
 Engineering
Saranathan College of Engineering
Trichy, India

Subhasis Roy
Department of Chemical Engineering
University of Calcutta
Kolkata, India

A. Saravanan
Department of Production Engineering
National Institute of Technology
Trichy, India

S. Sathees Kumar
Department of Mechanical Engineering
CMR institute of Technology
Hyderabad, India

Nitisha Sharma
Department of Civil Engineering
Shoolini University
Himachal Pradesh, India

Parveen Sihag
Department of Civil Engineering
Shoolini University
Himachal Pradesh, India

Shubham Tayal
Department of Electronics
 and Communication Engineering
SR University
Warangal, India

Mohindra Singh Thakur
Department of Civil Engineering
Shoolini University
Himachal Pradesh, India

Laxman Raju Thoutam
Department of Electronics and
 Communications Engineering
SR University
Telangana, India

Ankita Upadhya
Department of Civil Engineering
Shoolini University
Himachal Pradesh, India

Sergij Vambol
Life Safety and Law Department
Kharkiv Petro Vasylenko National
 Technical University of Agriculture
Kharkiv, Ukraine

Viola Vambol
Educational and Scientific Department
 of Occupational Safety and Health
National Scientific and Research
 Institute of Industrial Safety
 and Occupational Safety and Health
Kyyiv, Ukraine

P. Vivekanandhan
Department of Metallurgical
 and Materials Engineering
National Institute of Technology
Tiruchirapalli, India

1 Fabrication and Characterization of Materials

Laxman Raju Thoutam
SR University

CONTENTS

1.1 Introduction 1
1.2 Material Synthesis Techniques 2
    1.2.1 Czochralski Technique 3
    1.2.2 Chemical Vapor Transport Technique 4
    1.2.3 Physical Vapor Deposition 5
        1.2.3.1 Sputtering 7
        1.2.3.2 Pulsed Laser Deposition 8
1.3 Nanofabrication 9
    1.3.1 Photolithography 10
    1.3.2 Electron-Beam Lithography 11
1.4 Materials Characterization 13
    1.4.1 Scanning Electron Microscope 13
    1.4.2 Transmission Electron Microscope 15
    1.4.3 Scanning Tunneling Microscope 16
References 17

1.1 INTRODUCTION

The impact of materials science is ubiquitous in all branches of science and engineering. The study of matter from nanoscopic detail (atomic scale) to macroscopic detail (millimeter scale) helps to explore the myriad ways in which different chemical elements can be combined to form unique materials with diverse physical and chemical properties. The change in material-processing technologies over the last century has made great inroads into the development of modern electronics with faster processing speeds; to improve safety and gas mileage in the automobile industry; in the medical diagnosis of life-threatening diseases; and in increasing the efficiency of renewable energy-harvesting devices. Past few decade's long valiant efforts in exploring new material-processing techniques that include casting, additive manufacturing, electrochemical, coating, and vapor deposition techniques have shown fruitful long-term results in advanced material synthesis methods. Advances in recent material synthesis techniques have led to the growth of many novel material systems that include perovskite oxides, topological insulators, Weyl semimetals, and two-dimensional chalcogenides that exhibit exotic properties with unique applications. The numerous directions in advanced characterization techniques have evolved in the recent decade that enabled us to explore novel and exotic quantum phenomena. Recent technological developments in optical and electronic instrumentation have enabled the researchers to probe materials at an atomic level to qualify and quantify lattice constants, nature and bonding of atoms, defect density, and interface and surface quality with high precision that aids in optimizing the growth conditions to synthesize "*high-quality materials.*" The ability to synthesize and characterize a diverse set of materials using modern tools has impacted science and society in many profound ways. The focus of this chapter is to introduce various fabrication techniques that are commonly used to synthesize different material systems and characterize the same using modern advanced tools to study their structure–property relationship. Please note that this chapter presents an overview of various fabrication and characterization techniques to familiarize the readers with fundamental concepts. The readers are referred to in-depth studies related to different fabrication and characterization techniques for a more detailed study, as such topics are beyond the scope of this chapter.

1.2 MATERIAL SYNTHESIS TECHNIQUES

The synthesis of materials on a large scale is challenging in many ways, and it is of utmost importance to maintain a proper stoichiometric ratio of different chemical elements during the growth process. The presence of non-stoichiometry yields the formation of voids[1] and defects[2] and in some cases leads to the formation of secondary phases[3] that degrade the performance of the material being synthesized.[4] For example, all modern electronics rely on silicon for chip manufacturing, and silicon when exposed to oxygen forms silicon dioxide (SiO_2), a perfect insulator material with a high bandgap of 8.9 eV,[5] and acts as a passivation layer in its pristine state. However, a slight deviation from the nominal stoichiometry of $SiO_{2\pm x}$ will result in poor interface properties that might affect the performance of the chip. Note that the performance criteria of a material are subjective to its application.

Crystal growth methods differ from each other in terms of their starting raw material type, viz., melt, solid, vapor, and solution. The fundamental physical property, melting point, of a material plays a major role in choosing the raw material as it dictates the state of the material, viz., solid, liquid, and vapor states. Melting point can be assumed as the strength of a crystal lattice that depends on intermolecular forces, molecular symmetry, and conformational degrees of freedom of a material.[6] If sufficient energy is supplied to a material beyond its melting point, it changes from one form of matter to the other based on its thermodynamic conditions, namely pressure, volume, and density of the material. The technique to be used to synthesize a material depends on the type of material, cost of the growth process, reliability of the process, and quantity and quality of the synthesized material. This chapter discusses some commonly utilized techniques for material synthesis.

1.2.1 Czochralski Technique

This is the most common growth technique to make bulk large single crystals of materials, invented by the Polish scientist J. Czochralski in 1916.[7] This method is widely used to obtain large silicon ingots with a diameter of 15 mm and a weight exceeding 250 kg for making integrated chips in the semiconductor industry.[7] The technique involves heating of a solid, pure raw material that is placed inside a crucible above its melting point to form a melt. The crucible material in which the feed is placed should be able to withstand high temperatures. The crucible material should be chemically inert with the feed material at all times during the entire growth process to make sure the grown material is contaminant free. The entire growth set-up positioned on a rotating shaft is then placed in a vacuum chamber to avoid any contamination and heated above the melting point of the feed material. A schematic of the Czochralski growth technique is shown in Figure 1.1. A high purity seed crystal (same material as that to be grown) is placed in another rotating shaft and introduced from the top into the melt and is slowly pulled upward. Note that the rotation of the crucible and that of the seed crystal are in opposite directions. As the seed crystal is pulled atop, surface tension causes the melt near the seed crystal to rise above its uniform level to catch up with the withdrawing seed crystal, and in the process, a small single crystallite forms at the interface of the melt and the seed crystal. This process continues as the rotating seed crystal is slowly pulled away from the melt resulting in the formation of a large single crystal ingot. As the growth process is nearing its end, the pull rate of the seed crystal is slowly released to have a reduced diameter at the interface between the melt and the crystal, and eventually, the crystal ends with

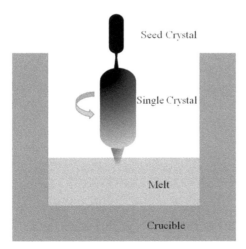

FIGURE 1.1 Schematic of the Czochralski growth technique showing the crucible, melt, and seed crystal only. The entire set-up is enclosed in a vacuum chamber and is surrounded by a heater set-up to supply thermal energy to the melt during the growth process. Please note that the rotation of the seed crystal is shown and the reader is advised to know that the crucible also rotates albeit in different directions.

a cone-like structure, much like the seed crystal shape. This process of increasing the pull rate near the growth end time ensures that no deformations are formed in the crystal and it is free of any defects.[7]

The growth conditions that primarily affect the quality of single crystal material are the pulling speed of the seed crystal, rotational speeds of the seed crystal and the crucible, and heating power of the melt.[7] The pull rate of the seed crystal determines the quality of the grown crystal; the slower the pull rate, the higher the quality of the crystal and the slower the growth rate. The length and the crystalline nature of the obtained single crystal material depend on the amount of feed material placed in the crucible and the crystal orientation of the seed crystal, respectively. This method can be employed to synthesize a wide range of different material systems in bulk quantity, provided a high purity seed crystal with the same chemical composition is available.

1.2.2 Chemical Vapor Transport Technique

The chemical vapor transport (CVT) technique was developed by Schäfer and has been used to synthesize many material systems including basic elemental materials, intermetallic compounds, halides, oxides, sulfides, selenides, tellurides, phosphides, pnictides, and many more in both bulk and micro- and nanostructures.[8,9] This method involves transfer of a heated solid material at one end of an ampoule in the presence of a transport agent that converts the solid material into gaseous phase and deposits at the other end of the ampoule. A typical transport agent comprises halogens and their compounds, viz., iodine, bromine, chlorine, and hydrogen-based halides. The experimental set-up usually consists of a two-zone temperature furnace, the transport agent, and the precursor sealed in an ampoule as shown in Figure 1.2.

The critical parameters that affect the growth mechanism are growth temperature, type of transport agent, rate of mass transport, and the free energy of the reaction. Typically, the reaction rate is very slow and the growth reaction can even extend up to few weeks. The transport of the vapor from the source to the destination depends on the pressure inside the ampoule. If the pressure inside the ampoule is very small, the transport happens via a diffusion process. However, if the pressure inside the ampoule is high, the transport occurs through a convection process. Note that the

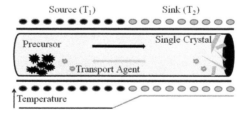

FIGURE 1.2 Schematic of the typical chemical vapor transport growth technique showing the precursor, transport agent, and the single crystal at the left, middle, and right-hand side of the sealed ampoule, respectively. A temperature gradient is set up across the ampoule to facilitate the growth process, and the temperature gradient is not unique and changes with different material systems.

Material Fabrication and Characterization

length of the ampoule is critical as it might favor any mixed-growth mechanisms, if it is too small. The overall chemical reaction inside the ampoule can be described using the following equation[9]:

$$iA_{(s,l)} + jB_{(g)} \rightleftharpoons kAB_{(g)} \qquad (1.1)$$

where i, j, and k correspond to stoichiometric coefficients in the transport equation. The raw material "A" can be in either solid or liquid state reacting with the transport agent "B"' at a particular temperature T_1 to form a vapor-phase intermediate product "AB." The vapor-phase intermediate product traverses the ampoule through a temperature gradient and undergoes a reversible reaction at a suitable temperature T_2 to disassociate into the transport agent "B" and the intended crystalline material "A." If the vapor partial pressure of the precursor is too low, a chemical substance can be added to increase the volatile nature of the raw material that favors gaseous transport from the source to the destination. However, if the volatility of the precursor is too high, the intermediate compound "AB" becomes highly stable, has a little probability of decomposition, and hinders the actual growth. For optimum growth, the appropriate stoichiometric proportions of the raw material and the transport agent must be selected. The temperature gradient across the ampoule should be uniform, and any non-uniform temperature distribution can result in re-evaporation of the single crystal preferably at higher temperatures. This technique majorly suffers from the presence of minute amounts of transport agent as impurities in the single crystal.

1.2.3 Physical Vapor Deposition

The physical vapor deposition (PVD) technique involves the process of directly converting a solid raw material into its vapor state and condensing it onto a temperature-controlled substance as a *thin film* inside a vacuum chamber. It can be employed to deposit metals, non-metals, insulators, oxides, ceramics, alloys, glasses, and polymers and is widely used in modern semiconductor industries. The role of the vacuum system in the PVD technique is two-fold. At first, it evacuates all the contaminants from the deposition work chamber, and on the other hand, it increases the mean free path of the deposited atoms/molecules. Mean free path is defined as the average distance traveled by an atom/molecule before a collision, and it is a function of temperature (T) and pressure (P), which is given by[10]

$$\lambda = \frac{kT}{\sqrt{2}\pi P d^2} \qquad (1.2)$$

where k is the Boltzmann constant and d is the diameter of the atom/molecule. The mean free path has a direct say over the film composition, and the deposition rate is a function of the mass flow rate between the source and the substance inside the work chamber. The maintenance of very low-pressure levels by the use of different vacuum pumps (that operate at different vacuum levels) ensures that the atoms/molecules travel farther distance before colliding with other atoms/molecules. For thermal evaporation, typically a pressure of 10^{-6} Torr is maintained in the work chamber that ensures a mean free path of 60 m, for a 4 Å atom/molecule. Note that in some PVD techniques (e.g., sputtering), reactive gases (argon and oxygen) will be introduced during deposition, and the background pressure will reach a level of around ~1 Torr, decreasing its mean free path to 60 µm for the same radius of atoms/molecules.

The material to be deposited is placed inside a crucible and is heated above its melting point using an external energy source. The type of energy source is different for different PVD techniques. For simple resistive evaporation, a highly pure raw material is placed on a tungsten/molybdenum filament, and the filament is heated to an extremely high temperature by sending current through the filament as shown in Figure 1.3. The current flowing through the filament controls the material evaporation and thus the deposition rate. It is important to note that the temperature profile over the entire heated filament is not uniform and this leads to non-uniformity of the evaporated materials. The simple evaporation technique also suffers from contamination issues, as any other material present on the filament can also get deposited onto the substrate during the deposition process. This technique is primarily used for depositing elemental materials and often becomes tedious to deposit composite materials. For composite material growth, the materials with different melting points reach the substrate at different times since the same heat energy is supplied to the composite material and this affects the stoichiometric composition of the deposited film.

The shortcomings of the simple resistive evaporation technique are addressed in electron-beam evaporation, in which a high-intensity controlled beam of electrons with an energy of around ~15 keV is focused on a source material that needs to be deposited. The electron energy melts the source material, evaporates it, and the vapor gets condensed onto the substrate, forming a thin film. The raw material is placed in a crucible and is water cooled to avoid any thermal gradients across the crucible that

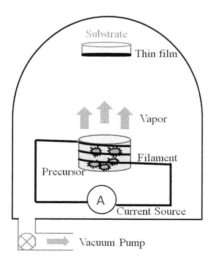

FIGURE 1.3 Schematic of the resistive evaporation growth technique showing the precursor, filament, current source, and vapor flow direction. The work chamber is maintained at a pressure of 10^{-6} Torr or less during deposition by using vacuum pumps (not shown here).

might lead to the damage of the crucible. The substrates are typically placed over the source material and are rotated to ensure uniform coverage of the film. The substrates are maintained at a higher temperature during the growth process to improve adhesion and for uniform coverage of the deposited material. The electron energy can be steered to different locations using dithered magnetic fields to evaporate different materials that facilitate the composite material/alloy deposition. Since the electron energy source behaves like a point-like source, shadowing effects are of concern that can result in non-uniformity of the film coverage during deposition. A special set-up of the substrate holder that includes rotating out-of-plane planetary architecture may eliminate the shadowing problem during deposition. A consequence of using high energetic electrons to bombard the target material is the generation of X-rays and secondary electrons that might damage the substrate and the deposited film.[10]

1.2.3.1 Sputtering

Sputtering is a PVD method in which target atoms are ejected out by bombardment with high energetic ions. The energy is transferred from incoming ions to target atoms to break the bonds holding the atom onto the solid. Sputtering yields a better step coverage than evaporation and less radiation damage, and is good at producing layers of compound materials and alloys, thus making it the most desirable technique for most of the silicon-based technologies. A simple sputtering model can be described by using a parallel plate plasma reactor in a vacuum chamber. The target material (to be deposited) is connected to a negative voltage terminal, and the substrate facing the target is mounted to the anode terminal, as shown in Figure 1.4. The voltages applied to the electrodes are in the range of 1–5 KeV that creates electric plasma in the deposition chamber.[11] The externally introduced positive neutral gaseous atom (usually argon) strikes the target surface and knocks off the neutral target atoms. These neutral target atoms travel further toward the anode (substrate) and are deposited as a thin film on the substrate. There will be some target atoms ionized,

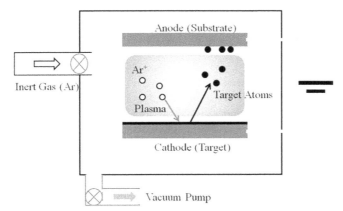

FIGURE 1.4 Schematic of the sputtering growth technique. The neutral argon atoms are ionized by the plasma and move toward the cathode terminal and bombard with the target material knocking off atoms. These knocked-off atoms move toward the anode and get deposited on the substrate near the anode terminal.

and these ionized ions interact with the gaseous atoms enabling continuity of the plasma within the chamber.

The deposition rate can be varied by changing the number of gaseous argon atoms inside the chamber and bias voltages applied to the electrodes. The spacing between anode and cathode must be optimum so as to collect all of the ejected ions from the target material and is usually less than 10 cm. This technique can be employed for composite materials and alloy depositions. Since the distance between the target and the substrate is very small, a uniform coverage of thin film is expected on the substrate. The electric plasma created during the growth process is in close proximity to the target material, and this induces thermal fatigue in the target and the composition of the target material may degrade during the aging process.

1.2.3.2 Pulsed Laser Deposition

Pulsed laser deposition (PLD) utilizes a high-power laser to ablate the target material to deposit on the substrate positioned within the chamber. Ablation is a process of removal of a material from the target species by vaporization. The source of the laser is outside the vacuum chamber and thus avoids any traces of contamination into the chamber, unlike other deposition systems where the source is positioned within the vacuum chamber. The film growth on the substrate is further enhanced by the introduction of a reactive gas into the chamber during deposition.

The heart of the PLD system is the use of lasers. The lasers employed in the PLD system have a general wavelength of the range of 200–400 nm with very short pulse durations as low as 10 ns. The range of 200–400 nm generates laser sources that can deliver a high power density of more than 1 joule/cm^2 to the target species at any given point of time. Beam quality and coherent characteristics of the laser output effectively define the stoichiometric compositions of the deposited films. A laser source is always outside of the deposition chamber and is made to pass through optical lenses before entering the chamber. The optical lenses must be always clean of any contamination as they might degrade the energy of the incoming laser beam, affecting the quality of the deposited material. With the PLD technique, deposition of multiple target species within a single run on the same substrate to have different layers of the materials is carried out sequentially to synthesize complex thin films and heterostructures.

PLD deposition is a sequential process wherein (i) laser ablate the target material to form a plume; (ii) the plume is expanded in the forward direction toward the substrate; and (iii) the plume gets deposited on the substrate forming a thin film. The first step is assuming that the laser is completely warmed up and pulsing at a specified rate onto a target. When the energized laser beam is incident on the target, the entire surface under the incident laser beam is vaporized within anywhere from 1 ns to 1 ps. The amount of material that can be vaporized is dependent on the wavelength of the laser beam and the index of refraction of the target. The surface of the target is then locally heated up to a very hot temperature in a short amount of time, anywhere from 8000 to 10,000 K.[12] An observable laser plume is seen moving toward the substrate in the form of a cone as shown in Figure 1.5.

The second step involves the plasma actually growing and extending out to the substrate for deposition. The third step is when the ablated target particles gets

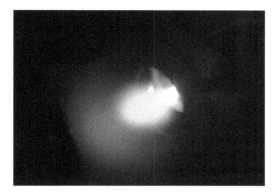

FIGURE 1.5 Picture of the laser plume taken from the Neocera Pulsed Laser Deposition System at the Microelectronics Research and Development Laboratory at Northern Illinois University. The dark intense spot is at the target (right-hand side of the image), and the gray Gaussian-like distribution is the plume traveling toward the substrate (left-hand side of the image).

deposited on the substrate as a thin film. The particles inside the plume are very highly energized, and when they collide with the substrate, they may do one of two things. They might crash into the substrate making tiny micro-cracks and even removing the top few layers of the substrate. The other thing that might happen is when the substrate is heated to a certain temperature, that increases the probability of a high-energized target particle to deposit onto the substrate, matching the thermal equilibrium. The deposition rate of the films is controlled by varying the lasing energy, altering the physical distance between the target and the substrate, and the background pressure inside the chamber. The unique advantage of PLD is that a complex material can be easily ablated due to the presence of high lasing energy concentrated on to a small spot size on the target. It is suggested to rotate the target as laser impinges on it, since lasing at the same spot will create a depression in the target and thus will affect future depositions.

1.3 NANOFABRICATION

The fast pace increase in technology and instrumentation in the fields of science and engineering facilitates the making of modern integrated circuits smaller in size and efficient in operation. An integrated circuit consists of active and passive components that are interconnected in a prerequisite manner to perform a specialized task. Modern IC fabrication techniques focus on to develop small sized wafers (~nanometer scale) at the expense of complex design architectures. The previously discussed material growth techniques are planar and form continuous films. The interconnections between different materials and different layers on a given wafer seemingly become hard. For example, insulators are used to separate two conductive materials for electrical isolation purposes; donor atoms are implanted into selective regions of intrinsic silicon to make the transistor architecture. This can be accomplished by using the lithography technique. In simple terms, lithography is defined as transfer of various geometric shapes on a mask onto a smooth planar surface, usually a thin film. This technique is most commonly used to define, design, and transfer different patterns on to the wafer.

1.3.1 Photolithography

Photolithography uses ultraviolet (UV) rays as a light source to transfer the patterns from the mask onto the wafer. Photolithography encompasses a whole range of steps to transfer the defined patterns onto the wafer. The entire process is done inside a special laboratory called *"Cleanroom,"* which at all times maintains and controls the air particulates present in a given square meter, temperature, pressure, and humidity of the contained space. The presence of minute contaminations or any dust particulates either on the mask or on the substrate yields the formation of defects that ruin the entire integrated circuit production process.

Photolithography involves many sequential steps and are (i) surface cleaning and pre-cleaning of the substrates before the photoresist coating process. Typical contaminants include atmospheric dust, dust from wafer scribing and slicing, and photoresist residues from previous photographic steps. The aforementioned dust particulates can be removed by putting the wafers in an ultrasonic bath of organic solvents like acetone and isopropyl alcohol. (ii) Spin coating the wafer with a uniform layer of sacrificial photoresist layer. A photoresist is a light-sensitive material and is made up of organic molecules. Spin coating is achieved by putting the wafer on a vacuum chuck and spinning it at a speed of 3000–5000 rpm. The photoresist when exposed to UV rays becomes soluble in a solution. (iii) Soft baking or prebaking the wafer for sufficient time to improve the adhesion of the photoresist and wafer; it also removes the coating solvent from the photoresist. (iv) Mask alignment is the stage at which a predefined geometrical design patterned with a photographic emulsion on a glass plate is placed over the wafer to transfer the intended design. The pre-patterned glass plate should possess alignment marks that serve as reference pointers for mapping wafer and the mask for each individual design layer. (v) Photoresist exposure involves the exposure of the wafer and mask combination to external UV radiation. The UV light cross-links the polymers present in the photoresist and makes them soluble/insoluble in the solution. Two types of photoresists, viz., positive and negative photoresists, exist. For positive photoresists, the area(s) where it is exposed to UV light becomes soluble in the solution and comes off during the developer stage. For negative photoresists, UV light cross-links the polymer and the exposed area remains intact when put in the developer solution. The process flow of photolithography using two types of resist is depicted in Figure 1.6.

After the exposure, the desired geometrical pattern is transferred onto the wafer and follows the etching process, to remove any barrier material not protected by the hardened photoresist. This process repeats for different geometrical designs and multiple layers of an integrated circuit fabrication.

The complete photolithography process (five steps mentioned above) is characterized by the smallest (finest resolution) lines or windows that can be produced on a wafer and is called as the minimum line width or minimum feature size. The minimum feature size is a metric of the type of exposure system employed and process

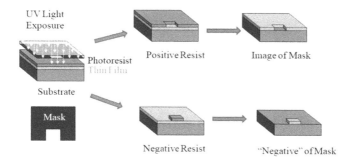

FIGURE 1.6 The schematic representation of working of a positive and a negative photoresist when exposed to UV light using the same mask. For the positive photoresist, the image of the mask is present on the wafer. For the negative photoresist, exposed areas remain, which is the reason for the negative image of the mask.

parameters, and the major limitation of optical photolithography is the fundamental consequence of light diffraction. For a single-wavelength projection photolithography system, the minimum feature size is given by the Rayleigh criterion[10]:

$$w_{min} = F = k\frac{\lambda}{\text{NA}} \qquad (1.3)$$

where λ is the wavelength, NA is the numerical aperture, a measure of the light-collecting power of the projection lens, and k depends on the photoresist properties and the "quality" of the optical system. NA for the best projection system is less than 0.8, and the theoretical lowest limit of k is 0.25. This puts a stringent hard limit on the minimum feature size of around 1–2 μm for all practical applications using optical photolithography. However, this can be further reduced by increasing NA, which comes with a price. The increase of NA decreases the depth of the focus of the projection system and is defined as the distance that the wafer can be moved relative to (closer to or farther from) the projection lens and still keep the image in focus on the wafer. The decrease in the depth of focus limits the physical movement of the wafer relative to the mask in the vertical direction, and in some cases, it leads to scratches on the mask that will deteriorate the efficacy of the mask, if it is too close. Apart from the minimum feature size restricted to micron level, another major disadvantage of optical photolithography is the use of a separate mask for each design layer and the number of masks scales with the complexity of the integrated circuit.

1.3.2 Electron-Beam Lithography

Electron-beam lithography (EBL) is an advanced and sophisticated technique, which uses electrons to define geometrical patterns onto the wafers at a nanometer scale. The EBL technique is a dynamic technique, where the electrons with much smaller wavelength (few nanometers depending on energy) achieve a greater resolution, with a maximum resolution around 1–10 nm depending on the process parameters. The electrons are directly focused onto the electron-beam (E-beam) resists to define and transfer the feature size on the wafer, and thus it is called as the direct-write technology. EBL eliminates the use of any intermediate mask. The EBL technique can transfer complex design patterns directly onto wafer via scanning mode. The scanning mode increases the design write-time and puts a stringent limit on the usage of EBL for large-scale productions.

The typical experimental set-up presented in Figure 1.7 is housed inside a vacuum chamber and has an electron source at the top of the column. The electrons can be emitted from a conductive tip either by heating it to an extremely high temperature (thermionic emission) or by applying a strong electric field between the conductive tip and an anode (field emission). The detailed description and the difference between thermionic and field emission are outside the scope of this chapter and the reader is advised to read the literature for a complete review. The electron spread across the column is limited by a pair of condenser lenses that restrict the electron flow to be focused. The condenser lens work much like an optical lens, except for the fact that they work on the principle of Lorentz force $F = q(\vec{E} + v \times \vec{B})$, i.e., the direction of the charged particle (electron) inside the column can be controlled by the appropriate electric and magnetic field values applied to condenser lens. The condenser lens is primarily used to form the electron beam and control the diameter (spot size) and current flow of the electron beam. The aperture is a small hole through which the electrons pass down the column. The aperture helps to blank off the electron beam when not writing and block off any contaminant flow down the column during the writing process. The beam blanker helps to deflect the electron beam off the wafer and is primarily used between the physical movements of the wafer during scanning mode. The beam deflector helps to scan the surface of the wafer at different locations

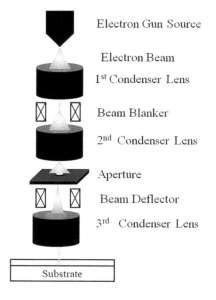

FIGURE 1.7 Schematic image of a typical electron-beam lithography system, adapted from Ref. [13], shows the electron path as it passes through different sections. Note that the entire set-up is enclosed in a vacuum chamber and is positioned on a vibration-free table (not shown here).

during the direct-write process. The entire set-up is placed inside a vacuum chamber as electrons can easily dither by the air particulates or other contaminants during their travel from the source to the wafer. The complex circuitry and the fact that it is very expensive limit the usage of EBL to research purposes only, and its application in the industrial sector is rather limited to mask making for optical lithography procedures.

1.4 MATERIALS CHARACTERIZATION

The characterization of materials using a variety of techniques will reveal the physical and chemical properties. The different techniques that probe electronic, optical, magnetic, mechanical, chemical, and structural properties will reveal the fundamental understanding of a material. The electronic, magnetic, and thermal properties directly depend on the crystal structure of the material. Thus, material characterization can help to map the structure–property relationship of a material. Note that the probing techniques can be destructive, meaning they can destroy the pristine nature of the material, and thus a detailed overview of a particular characterization technique should be known prior to probing. The increase of technology and instrumentation in the field of science and engineering has made a drastic change in the way we characterized the materials in the last decade. Many new and novel physical phenomena have been experimentally verified with the use of advanced characterization techniques. For example, looking at an electron waveform inside an atom, the position of individual atoms in a complex material and how different atoms bond together at the interface were experimentally visualized in recent years.[14] Selection of a characterization technique depends on the type of material and the property to be probed. This chapter tries to overview some generic characterization techniques that employ electron probes for material characterization.

1.4.1 SCANNING ELECTRON MICROSCOPE

A scanning electron microscope (SEM) uses electrons as a probe to image a sample. The fact that electrons are used as probe elements yields a unique advantage of having a very small wavelength (wavelength depends on energy: $E = h\upsilon$) that can reveal atomic-level detail that can help to understand the material characteristics at the nanoscale. The SEM has a large depth of field, which allows a large amount of the sample to be focused at one time and produces a three-dimensional image of the sample. The SEM produces images of high resolution that aid to probe nanoscale features at a high magnification. The combination of higher magnification, larger depth of field, greater resolution, and compositional information makes the SEM one of the versatile instruments used for material characterization. The experimental set-up of SEM looks exactly similar to EBL as depicted in Figure 1.7. In fact, all modern EBLs have the capability to perform scanning electron microscopy. A SEM uses a point-to-point measurement strategy to image a sample material. A high energy electron beam is used to excite the specimen, and the signals are collected and analyzed so that an image can be constructed. The signals carry topological (surface features), morphology (size and shape), chemical (elemental composition and relative atomic ratios), and crystallographic information (grain size and boundaries), respectively, of the material surface. The interaction between the electrons and the solid surface is not trivial and is complicated due to the charge of the electron, subsequent interactions between the solid and the electron that depends on the penetration depth of the electrons into the sample, which is a function of energy of the electrons, thickness, and the type of material under consideration.

The various interactions at the interface between incoming electrons and a sample are represented in Figure 1.8. The secondary electrons are produced by inelastic interactions of high-energy electrons with loosely bound valence electrons at the surface of the specimen, causing the ejection of the electrons from the atoms. These ejected electrons with an energy around 10–50 eV are termed "secondary electrons."[15] Each incident electron can produce several secondary electrons. Due to their low energy, only secondary electrons that are very near the surface (<10 nm) can exit the sample and be examined to reveal the topographical information of the sample. On the other hand, back-scattered electrons are produced by elastic interactions of beam electrons with nuclei of atoms in the specimen, and they have high energy and large escape depth. Due to the direct interaction with the nucleus of atoms, the image shows characteristics of atomic number contrast.

The high energetic incident electrons are also responsible for the generation of characteristic X-rays that are specific to the material being characterized. The analysis of the intensities and energy of the same will reveal compositional information about the material. The major drawback of the SEM is that it cannot directly image the insulating samples. The insulating sample gets charged up due to incident electrons, and this extra charge at the interface will result in poor quality of images. A few layers of metallic coating are deposited on the insulating material to avoid charge buildup and facilitate good imaging, and note that the thickness of the metallic coating should not interfere with the features of the insulating sample.

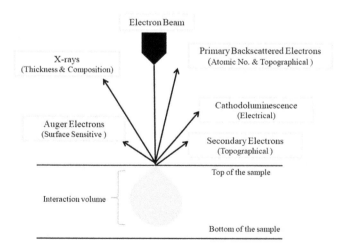

FIGURE 1.8 The list of possible interactions between the incoming electrons and the sample surface inside an SEM, adapted from Ref. [15].

1.4.2 Transmission Electron Microscope

A transmission electron microscope (TEM) uses electrons to probe the structural and compositional properties of the material at the atomic level. In a TEM, electrons are made to pass through the sample and the transmitted or diffracted data are analyzed for material characterization, unlike an SEM where the scattered electrons are analyzed. TEM uses very high energy electrons typically (>50 keV), which leads to a very smaller wavelength that can interact at the lattice level of the atom/molecule to reveal its properties with the highest possible resolution achieved by any other microscopy technique. The experimental set-up of TEM is similar to SEM and requires a vacuum level of 10^{-7} Torr. TEM uses a pair of projector lens to magnify the image or diffraction pattern onto the final fluorescent screen for visual display. TEM has the capability to collect the diffracted data of the material to reveal its crystalline structural properties. Due to its high resolution, it is more suitable for the structural characterization of nanomaterials to reveal their atomic structure, dislocation, and defect density. The sample preparation for TEM requires special attention as the sample has to be thin enough for the incident electrons to pass through. If the sample is thick, incident electrons undergo multiple scatterings and no transmitted electrons are available for imaging. Alternatively, the sample can be made in powder form, which is usually placed on a copper metal grid for imaging.

The diffracted electrons are analyzed using Bragg's law, $\lambda = 2d \sin\theta$, where λ is the incident wavelength of the energy, d is the lattice constant, and θ is the glancing angle. For a given incident electron intensity of energy $E \approx 200$ kV, the wavelength of electrons yields 0.0251 Å, and assuming a lattice constant of 2.5 Å results in a glancing angle of 0.288°.

Figure 1.9 shows a simple illustration of TEM-diffracted data that reveals the orientation of grain structure within the sample when exposed to the same incident beam of electrons. If the grain structure is normal to the sample surface, only one dark spot (Figure 1.9b) is present on the fluorescent screen, and if the grain structures are off-normal axis to the incident beam, the number and positions of the dark spots

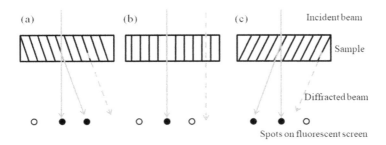

FIGURE 1.9 A sample schematic illustration of TEM-diffracted data of a sample with different grain structures. Note how the spots on the fluorescent screen reflect the true nature of the grain structure. (a) The dashed lines with no corresponding spots on fluorescent screens are used to illustrate the difference between the bright-field (b) and the dark-field (c) imaging techniques.

vary accordingly, revealing the grain structure of the sample, as shown in Figure 1.9a and c. There might be a difference in the contrast of diffracted spots on the fluorescent screen due to the intensity of the different grain structures and the presence of any defects in the sample. In TEM, two modes of imaging are possible viz., bright-field and dark-field imaging.[15] Bright-field imaging corresponds to unscattered (or transmitted) electron data. The elements with high crystalline mass appear dark in bright-field imaging mode. Dark-field imaging corresponds to scattered or diffracted data. The bright spots in the dark image reveal the presence of a crystalline material. The selection of the transmitted and diffracted data can be achieved by using an aperture in the TEM column.

1.4.3 SCANNING TUNNELING MICROSCOPE

The scanning tunneling microscope (STM) technique works on the principle of quantum tunneling. In quantum mechanics, when a particle hits a potential that does not have enough energy to pass (say inside the square well), the electron wave function dies off exponentially. If the well is short enough, there will be a finite probability of finding the particle on the other side of the potential, resulting in a tunneling current. For quantum tunneling to happen, the materials must be in close proximity to each other with the distance between the two less than a nanometer (<1 nm).[16] Scanning tunneling microscopy is an electron microscopy technique with a single atomic conductive tip that uses the quantum tunneling principle to image the sample with atomic resolution. The atomic tip is held within a nanometer distance above the sample surface. Due to close proximity, a tunneling current is established between the probe tip and the sample surface. Since both the tip and sample are conductive, electrons are present on both sides, and when they are brought close to each other separated by a narrow gap (<1 nm), they need to overcome a finite barrier to travel from the tip to the surface or vice versa. If the tip and the sample are held at the same electrical potential, their Fermi levels line up exactly and there are no empty states on either side available for tunneling. However, if an external voltage is applied between the sample and the tip, electrons start to tunnel from the sample to the tip, resulting in an electric signal. The tunneling current is a strong function of the distance between the atomic tip and the surface and it varies exponentially with distance. A very minute change in the distance can lead to a large change in tunneling currents. If the tip is slowly scanned across the sample, the tip is raised and lowered to keep the signal constant and yet maintain the distance. This enables the tip to follow the sample surface with atomic-level precision yielding topographical information. STM can be operated in two modes viz., constant height mode and constant current mode. In constant height mode, the separation between the tip and the surface is kept constant, and the tunneling current is a function of the lateral position of the tip that reveals the surface image. This mode is more suitable for atomic flat surfaces, and any irregularities on the sample surface may crash the tip. In constant current mode, the tunneling current is kept constant by constantly adjusting the tip height using the piezoelectric crystals. The change is piezovoltage, while scanning directly correlates the topography of the sample surface.

STM uses feedback servo loops and converses piezoelectricity to maintain angstrom level separation between the tip and the sample surface. The application of an electric field to a piezoelectric crystal makes it distort (converse piezoelectricity effect). The distortions of a piezocrystal are usually on the order of micrometers, which is in the scale needed to keep the tip of the STM a couple of Angstroms from the surface. STM is used to image surface roughness of many electronic and optical components. The variations in the tunneling current, voltage, and separation between the conductive tip and the sample surface will help to uncover the electronic properties of different materials. Please note that STM works for conducting samples only. An STM measures the electron density clouds on the sample surface and does not yield any direct information about the nuclear position. Since STM works on a surface level of a sample, any physical contamination or adsorbent on the sample should be avoided. The angstrom level separation between the tip and the surface is quite sensitive to the external environment, and thus, ultrahigh vacuum conditions are maintained during the characterization and the entire set-up is positioned on a vibration isolation system.

REFERENCES

1. S. Anderson, "*Defect Chemistry and Non-Stoichiometric Compounds*" Modern Aspects of Solid State Chemistry. Springer: Boston, 1970.
2. T. Wang, L. R. Thoutam, A. Prakash, W. Nunn, G. Haugstad, and B. Jalan, "Defect-driven localization crossovers in MBE-grown La-doped $SrSnO_3$ films" *Phys. Rev. Mater. (Rapid Communication)*, 1, 061601(R), 2017.
3. S. Susan, G. Galina, G. Maxim, D. Mirjana, P.R. Alejandro, I. Victor, S. Laudia, K. Juran, J. William, and M. Manuel, "Point defects, compositional fluctuations, and secondary phases in non-stoichiometric kesterites" *J. Phys. Energy*, 2, 012002, 2020.
4. Gong, J. Yang, M. Ge, Y. Wang, D. Liang, L. Luo, X. Yan, W. Zhen, S. Weng, and L. Pi, "Non-stoichiometry effects on the extreme magnetoresistance in Weyl semimetal WTe2" *Chinese Phys. Lett.*, 35, 097101, 2018.
5. M. Ravindra, and J. Narayan, "Optical properties of amorphous silicon and silicon dioxide" *J. Appl. Phys.*, 60, 1139, 1986.
6. C. Dearden, "The QSAR prediction of melting point, a property of environmental relevance" *Sci. Total. Environ.*, 59, 109, 1991.
7. W. Zulehner, "Historical overview of silicon crystal pulling development" *Mat. Sci. Eng. B*, 73, 7. H. Schafer, *Chemische Transportreaktionen*, VCH: Weinheim, 1962.
8. H. Schafer, "Der Chemische Transport und die" "Löslichkeit des Bodenkörpers in der Gasphase"" *Z. Anorg. Allg. Chem.*, 400, 242–252, 1973.
9. M. Binnewies, R. Glaum, M. Schmidt, and P. Schmidt, "Chemical vapor transport reactions – A historical review" *Z. Anorg. Allg. Chem.*, 10, 1002, 2013.
10. R.C. Jaeger, "*Introduction to Microelectronic Fabrication*" 2nd edition, Prentice Hall: New Jersey, 2002.
11. S. A. Campbell, "*The Science and Engineering of Microelectronic Fabrication*" 2nd edition, Oxford University Press: New York, 2001.
12. D.B. Chrisey, and G.H. Hubler, "*Pulsed Laser Deposition of Thin Films*", John Wiley & Sons: New York, 1994.
13. A. Pimpin, and W. Srituravanich, "Review of micro- and nanolithography techniques and their applications" *Eng. J.*, 17, 37, 2011.

14. R. Bach, D. Pope, S. Liou, and H. Batelaan, "Controlled double-slit electron diffraction" *New. J. Phys.* 15, 033108, 2013.
15. S.K. Kulkarni, *"Nanotechnology; Principles and Practices"* 3rd edition, Capital Publishing Company: New Delhi, 2015.
16. Binnig, and H. Rohrer, "Scanning tunneling microscopy" *IBM J. Res. Develop.*, 30, 355, 1986.

2 Application to Advanced Materials Simulation

Soumyajit Maitra, Souhardya Bera, and Subhasis Roy
University of Calcutta

CONTENTS

2.1 Introduction 19
2.2 The Entrance of DFT into Chemistry... 21
    2.2.1 Functional Exchange-Correlation... 21
        2.2.1.1 Local Density Approximation (LDA)................................ 21
        2.2.1.2 Generalized Gradient Approximation [GGA] 21
        2.2.1.3 Meta-GGA 22
        2.2.1.4 Hybrid Functionals: Hartree–Fock Exchanges in Functional Development...23
        2.2.1.5 Recent Progress in Functionals...23
2.3 Application of DFT in Photovoltaics 26
    2.3.1 Assessing the Fundamental Properties of Semiconductors Used in Photocatalysts and Photovoltaic Systems...............................26
        2.3.1.1 Methodology 26
        2.3.1.2 Results and Discussions 28
    2.3.2 Investigated Efficiency of Photocatalysis in a Monolayer g-C_3N_4/CdS Heterostructure [89] 33
    2.3.3 Perceptions from a DFT+U Investigation on Interfacial Interactions, Charge Transport, and Synergic Mechanisms in $BiNbO_4$/MWO_4 (M = Cd and Zn) Heterostructures for Hydrogen Processing [96] 36
        2.3.3.1 Computational Details 36
        2.3.3.2 Results and Discussions 37
2.4 Conclusion and Future Prospects 40
Acknowledgment 41
References.. 41

2.1 INTRODUCTION

With the increasing population on earth, it is believed that there would be a global energy shortage, coupled with weighty environmental pollution concerns and rising CO_2 emissions, arising out due to the low efficiency of fossil fuels [1]. Extracting sunlight energy is, therefore, a significant task. Collecting energy from the sun is a significant challenge for sustainable progress, and harvesting solar energy is a significant source of clean hydrogen fuel [2,3]. Solar energy harvesting can meet the global energy shortages and reduce dependence on fossil fuels. All of this can be accomplished even though just 1% of the 3,850,000 exajoules of solar energy every year is harvested [4]. Various solar power harnessing devices have recently been developed to transform solar energy into a more usable shape of energy, such as electricity through photovoltaic systems [5], photocatalysis, and other applications [5], all of which have the potential to generate energy at low economic costs [6]. For being a green and efficient technology, semiconductor photocatalysis [7] was a promising future in solar energy harvesting and pollution reduction in 1972 [8]. Semiconductor photocatalysts like TiO_2, ZnO, $SrTiO_3$ have exhibited high photocatalytic performance [9-10]. Semiconductors have an excellent bandgap, which can be changed at our behest photocatalysts for the conversion of light to other forms [11]. Table 2.1 shows tabulated form of the different semiconductors and its applications.

But with the increasing demand for solar energy, systems need to be developed to store this energy. As a sustainable, efficient, and clean energy, hydrogen (H_2) energy provides an excellent option to both store solar energy and reduce dependence on fossil fuels, thereby reducing pollution [12,13]. Thus, water, having a high hydrogen content, becomes the automatic choice to extract hydrogen [14]. But, the success of using hydrogen as energy depends on the production capacity of hydrogen. The hydrogen evolution reaction (HER) is a catalyst-activated process that needs a high activation energy of 238 KJ/mol to split water [15]. Due to their inherent stability and low-cost synthesis, metal oxides have become common photocatalyst materials [16].

The objective is to briefly demonstrate the application of density functional theory (DFT) in unraveling the intricate underlying electronic processes in photocatalysis from previous reports.

Currently, DFT is employed to calculate the structural, hydrogen evolution reaction (HER) mechanism, carrier separation and mobility, synergistic effect, and electronic and optical properties of these hybrid heterostructures. Instead of performing an exhaustive solution structure scan, DFT allows creating a collection of unique functionals and accessing their significance. The consistency of the feature used to

TABLE 2.1
Semiconductors and Applications

Applications	Examples of Applications
Inorganic photovoltaics	Si [126], Ge [127], GaAs [126][128], CdTe [126][129], $CuInSe_2$ [130], [131]
QD-based solar cells and DSSC	TiO_2, ZnO [132], SnO_2 [133], NiO [134], CdTe CdSe [135]
CO_2 generation	In $(OH)_3$ [136], TiO_2 [137, 138, 139]
Water splitting application	Fe_2O_3 [140], CdS [141], NiO [142], Bi_2S_3, Ta_3N_5 [143]
Waste photodegradation	$BiVO_4$ [144], Bi_2WO_6, TiO_2 [145]

define certain chosen properties, such as electronic exchange and correlation, will then be directly related to the accuracy of DFT.

2.2 THE ENTRANCE OF DFT INTO CHEMISTRY

Since the time of Thomas [17], Fermi [18], and Slater [19], DFT has had a very uneven past. Since each of the theories was based on certain assumptions, none were expected to provide accurate results. However, as the Hohenberg–Kohn theorem formulated, this fundamental change and the function of the density were used to obtain an exact quantum mechanical result. The Kohn–Sham (KS) equations quickly followed up, which proposed a more comprehensive and perfect expression based on non-interacting electrons, thereby consolidating DFT to well-acquainted concepts of orbitals in chemistry. The only problem, according to the Kohn–Sham paper, was to progress the construction of exchange-correlation functionals. It did, however, recommend that the local density application be used to understand the issue better. It soon caught the attention of solid-state physicists; however, integration into the chemistry culture took time, despite the fact that Slater's X theory had come before it.

2.2.1 Functional Exchange-Correlation

The chronology of development and application of exchange-correlation functional is critical to appreciating concepts in chemistry. Each new advance had its set of key achievements and limitations, and it is essential to understand the functions of modern DFT, so that it becomes easier to formulate more frameworks to work upon:

2.2.1.1 Local Density Approximation (LDA)

Dirac's equation was the first form known for the conversation of the electron gas uniformity [20]:

$$E_x^{\text{LDA}}[\rho] = -\frac{3}{4}\left(\frac{3}{\pi}\right)^{1/3} \int \rho^{4/3}\, dr \tag{2.1}$$

Monte Carlo's uniform gas simulations [21] parameterized the interpolations among the known structures in the low- and high-density limits instead of developing the functional from initial propositions. Perdew and Vosko et al. [22,23] have developed the presently employed correlation functionals of LDA.

2.2.1.2 Generalized Gradient Approximation [GGA]

While unchanging electron gas is a good example of a scheme that has even played a significant part in functional development, its density is very different from what is normally found by atomic or molecular methods. The Hohenberg–Kohn proof is an added generic equation and can be used to understand elementary information of the density gradient ($\nabla \rho$). In addition, there was confusion as to which analytical expressions [24] should be used to obtain the exact coefficients to be applied to each gradient expansion term. The reduced gradient dimensionless $x = |\nabla \rho|/\rho^{4/3}$ [or equivalent $2(3\pi^2)^{1/3}\ s = x$] used to express the gradient for the gradually changing uniform electron gas has the form

$$E_X^{\text{GEA}} = -\int \rho^{\frac{4}{3}} \left[\frac{3}{4}\left(\frac{3}{\pi}\right)^{\frac{1}{3}} + \frac{7x^2}{432\pi(3\pi^2)^{\frac{1}{3}}} + \cdots \right] dr \qquad (2.2)$$

However, the problem does not occur for regions where x is well defined (the nuclei); it is faced at places where $x \to \infty$ (atomic tails, where it has an exponentially decaying density). However, this is a useful example of the three main challenges in derivatives, solid-state physics problems, and complications in chemistry. The GGA, which denotes general gradient approximation, was, therefore, necessary and takes a simple form for exchange:

$$E_X^{\text{GGA}}[\rho, x] = \int \rho^{\frac{4}{3}} F(x) dr \qquad (2.3)$$

$F(x)$ is selected for the expansion gradient within the low x limits (Eq. 2.2). Several types of interchange functions are currently available as GGA. However, the two most used are:

$$E_X^{B88} = -\sum_{\sigma=\alpha,\beta} \int \rho\sigma^{\frac{4}{3}} \left[\frac{3}{4}\left(\frac{6}{\pi}\right)^{\frac{1}{3}} + \frac{\beta x_\sigma^2}{1 + 6\beta x_\sigma \sinh^{-1} x_\sigma} \right] dr \qquad (2.4)$$

$$E_X^{\text{PBE}} = -\int \rho^{\frac{4}{3}} \left[\frac{3}{4}\left(\frac{3}{\pi}\right)^{\frac{1}{3}} + \frac{\mu s^2}{1 + \frac{\mu s^2}{K}} \right] dr \qquad (2.5)$$

Correlation functional development is complex and consumes a longer time for LDA to develop. Coordinate scaling relations for connection [25,26] are difficult to use and thus are seldom applied. The two key functionals currently known are LYP [27] and PBE [48], but a number of other GGA functionals were evolved and have also been documented in the literature.

2.2.1.3 Meta-GGA

As the shift from LDA to the GGA was imminent, Perdew and Schmidt coined the term Meta-GGA, besides proposing Jacob's ladder of estimates [28] for the precise exchange functional correlation. This provides the next level of chemical accuracy that advances from Hartree world even without exchange-correlation (in

1 kcal/mol for energetics). The first rung begins as the LDA, followed by GGA, and the third rung as meta-GGA, which contains additional local ingredients in the form $E_{XC}^{MGGA} = \int \rho^{\frac{4}{3}} F\left(\rho(r), \nabla\rho(r), \nabla^2\rho(r), \tau(r)...\right) dr$ including the energy density (kinetic), $\tau = (1/2) \Sigma_i |3\phi i|^2$. There are now functionals that include the kinetic energy density, τ (which is often described without 1/2). Numerous functionals using τ, specifically correlation functionals are used on the basis of reasoning that one-orbital systems $\tau_\sigma = (3F\sigma)^2/8F\sigma$ can be used to remove $\sigma-\sigma$ correlation in one-orbital regions. The idea of functionals of these types is created from Becke like B88C or B95 or Becke-Roussel [29,30]. Many different functionals were also developed by Perdew et al., which includes TPSS [31] and many others with different ideas.

2.2.1.4 Hybrid Functionals: Hartree–Fock Exchanges in Functional Development

$$E_X^{HF} = -\frac{1}{2} \sum_{ij\sigma} \iint \frac{\varphi_{i\sigma}^*(r)\varphi_{j\sigma}(r)\varphi_{j\sigma}^*(r')\varphi_{i\sigma}(r')}{|r-r'|} dr\, dr' \qquad (2.6)$$

The original definition was taken from Axel Becke. He proposed to include some of them (Ex HF) in the functional system and proposed a linear model which embedded local DFA exchange and correlation-based functionals.

$$E_X^{BHH} = \frac{1}{2} E_X^{HF} + \frac{1}{2} W_1^{LDA} \qquad (2.7)$$

This has led to new features like BHLYP [32], which have not been equal to GGA functions. But ideas were quick to introduce some experimental knowledge to sharpen the idea and create

$$E_{xc}^{B3} = aE_x^{HF} + (1-a)E_x^{LDA} + b\Delta E_x^{B88} + cE_c^{GGA} + (1-c)E_c^{LDA} \qquad (2.8)$$

where a, b, and c are parameters selected to fit some sets of experimental data. This was administered into the Gaussian package [33] by

$$E_{xc}^{B3LYP} = 0.2E_x^{HF} + 0.8E_x^{LDA} + 0.72E_x^{B88} + 0.81E_c^{LYP} + 0.19E_c^{VWN} \qquad (2.9)$$

The VWN and LYP functionals are used (with some misperceptions about which parameterization (V or III) to apply). Following its success, it was commonly used in the way that a DFT calculation was seen analogous to the B3LYP computations. While the B3LYP calculations in chemistry have been very successful, the anticipations of DFT have been improved due to a certain number of error deductions.

2.2.1.5 Recent Progress in Functionals
Presently, there are some functionals developed that have had a consequential impact in the field of DFT.

a. **Range Separation:** Groups of Savin [34–36] and Gill [37,38] have come together to create a concept that allows the interaction between electron and electron to be divided into two parts – one short-range and one long-range. Because it is simple to estimate integrals on a Gaussian basis, the most widely used method is error function-based range separation.

$$\frac{1}{r_{ij}} = \frac{erf(\mu r_{ij})}{r_{ij}} + \frac{erfc(\mu r_{ij})}{r_{ij}} \qquad (2.10)$$

The Hartree–Fock long-range is specified by

$$E_x^{lr-HF} = -\frac{1}{2}\sum_{ij\sigma}\iint \frac{\varphi_{i\sigma}^*(r_1)\varphi_{j\sigma}(r_1)\varphi_{j\sigma}^* erf(\mu r_{12})\varphi_{j\sigma}^*(r_2)\varphi_{i\sigma}(r_2)}{|r_1 - r_2|} dr_1\, dr_2 \qquad (2.11)$$

The corresponding shape of the calculated long-range LDA exchange energy from the explicit form of the exchange hole [39] is

$$E_x^{lr-LDA} = \frac{1}{2}\sum_\sigma \int \rho_\sigma^{\frac{4}{3}} K_\sigma^{LDA}\left(\frac{8}{3}a_\sigma\left[\sqrt{\pi}erf\left(\frac{1}{2a_\sigma}\right) + 2a_\sigma(b_\sigma - c_\sigma)\right]\right) dr \qquad (2.12)$$

In atomic and molecular systems, the Hartree–Fock potential is correct at the asymptotic limit [40] in atomic and molecular systems. The following simple long-range corrected functional can be obtained if long-range Hartree–Fock can be intermixed with short-range:

$$E_{xc}^{LC-DFA} = E_X^{DFA} + \left(E_x^{lr-HF} - E_x^{lr-DFA}\right) + E_c^{DFA} \qquad (2.13)$$

While these functionals do not enhance thermochemistry science, they may greatly improve some properties like excitation energy [41].

b. **Fitting:** There are significant difficulties in terms of depth and complexity. So, it would be more realistic to work out forms from available experimental data, and that too with the help of few parameters. The three parameters in B3LYP fitted with the G1 experimental dataset proved very successful for the earliest development of functionals. Later Becke extended the idea in his B97 functional to ten parameters [42,43]. However, it is not clear how parameters are necessary for the exchange-correlation problem [44]. Still unknown, many functionals developed used a high-level parameterization, without a reasonable theoretical basis to whether so many are required actually to define the same. However, many exciting functionals, such as B98, VS98, and τ-HCTH [45,46], were developed. In many areas of chemistry, their performance improves significantly over the B3LYP-functional standard. The M06 range of functional features M06-L, M06-2X, M06, and M06-HF have created the best possible solutions [47,48].

c. **Connection Functionals: Adiabatic:** A fragment of Hartree–Fock was incorporated into the hybrid functional design, which was the final significant boost in functional development. This was justified by the adiabatic link [49]. This know-how on adiabatic connections can lead to the development of more efficient functional structures. The most superficial thought from Becke was to use a linear model to create functions using models for the adiabatic connection integrand

$$W_\lambda = a + b\lambda \tag{2.14}$$

based on the knowledge of the exact W_0 and W_1^{LDA}. Other ideas, such as building models using W_0, W_∞, W'_0, and W'_∞, employed mostly by Perdew and coworkers [50–52] can be used. Others involve [53] a form

$$W_\lambda = a + b\lambda/(1+c\lambda) \tag{2.15}$$

Teale et al. have recently conducted [54,55] some interesting research into the authentic adiabatic connections [56,57], based on Wu and Yang's theoretical formulation [58].

d. **Local Hybrids:** Some systems have to be quite different from the Hartree–Fock exchange. Defining a local version of specific exchanges at each point in space is another way to vary the amount of exact conversation.

$$e_X^{\sigma,\text{HF}}(r) = -\frac{1}{2}\sum_{ij}\int \frac{\varphi_{i\sigma}^*(r)\varphi_{j\sigma}(r)\varphi_{j\sigma}^*(r')\varphi_{i\sigma}(r')}{|r-r'|}dr' \tag{2.16}$$

With a local mixing function $a_\sigma(r)$, a local functional similar to B3LYP may be defined.

$$E_{xc}^{1H} = \sum_\sigma \int \left(a_\sigma(r)e_x^{\sigma,\text{HF}}(r) + (1-a_\sigma(r))e_x^{\sigma,\text{LDA}}(r)\right)dr + b\Delta E_x^{\text{GCA}} + E_c^{\text{GGA}} \tag{2.17}$$

These ideas have been used to develop many interesting functionals and ideas. However, some problems related to the choice of measurement in the development of hybrid functionals can also be completely overcome [59].

e. **Functionals Involving Unoccupied Orbitals and Eigenvalues:** The MP2 functional is the most straightforward functional using unoccupied orbitals and eigenvalues:

$$E_c^{\text{MP2}} = \frac{1}{4}\sum_{ij}^{occ}\sum_{ab}^{virt}\frac{\langle ij|ab\rangle^2}{\in_i + \in_j - \in_a - \in_b} \tag{2.18}$$

Here, eigenvalues presume a very important role. Where MP2 energy expression referring to the eigenvalues on the bottom is from Hartree–Fock, GL2 second-order correlation energy represents eigenvalues from a local potential [60]. The MP perturbation theory and the GL perturbation theory differ a lot from each other, although improvements are perceived in some cases [61]. For instance, it only worsens the situation when using the GL2 for simple systems [62]. Some well-fitted functional incorporating both MP2/GL2 like correlation term can be represented by [63]

$$E_{xc}^{\text{B2PLYP}} = aE_x^{\text{HF}} + (1-a)E_x^{\text{GGA}} + bE_c^{\text{GGA}} + (1-b)E_c^{\text{PT2}} \qquad (2.19)$$

The MP2 expression is referred to as PT2, but the eigenvalue at the bottom is a hybrid functional. The principle of the fluctuation-dissipation theorem is used in many newly developed functionals for unoccupied orbitals (derived from ancient concepts).

$$E_{xc}^{RPA} = -\frac{1}{2\pi} \int_0^1 \int_0^\infty \iint \frac{1}{r_{12}} \chi_{XC}^\lambda(r_1, r_2, i\omega) dr_1\, dr_2\, d\omega\, d\lambda \qquad (2.20)$$

Expressing the energy of the ground-state correlation over frequency is a type of energy excitation and leads to the approximation of the random phase (RPA)

$$\begin{pmatrix} A & B \\ -B & -A \end{pmatrix} \begin{pmatrix} X \\ Y \end{pmatrix} = \omega \begin{pmatrix} X \\ Y \end{pmatrix} \qquad (2.21)$$

where $A_{ia,bj} = (\varepsilon_i - \varepsilon_a)\delta_{ia,bj} + (ia|bj)$ and $B_{ia,jb} = (ia|jb)$. The RPA correlation energy is given by the difference between the many-body and single-particle excitations

$$E_c^{RPA} = \sum_{ia} \omega_{ia} - A_{ia,ia} \qquad (2.22)$$

Often referred to as the direct RPA, orbitals and eigenvalues like full RPA (or RPAX or RPAE) differ from other similar methods. Ideas also link RPA to the coupled-cluster formulation, which can help understand some of the methodological successes and shortcomings [64]. Additional developments include the use of range-separation [65,66] or extended formalisms, such as second-order screened exchange, SOSEX [67,68], to be associated with RPA.

2.3 APPLICATION OF DFT IN PHOTOVOLTAICS

2.3.1 ASSESSING THE FUNDAMENTAL PROPERTIES OF SEMICONDUCTORS USED IN PHOTOCATALYSTS AND PHOTOVOLTAIC SYSTEMS

2.3.1.1 Methodology

Here, different ways of computing other essential properties using DFT have been presented.

    a. **Bandgap Calculation:** The bandgap, E_g, is defined as the energy difference between the minimum energy of the conduction band and the maximum energy of the valence band. These energies are obtained through the self-consistently resolved monoelectronic Khon–Sham equations. Measurement resulted in the monoelectronic bandgap. A post-analysis can also be conducted on TD-DFT calculations to estimate the excitonic bandgap [69].

b. **Dielectric Constants:** Both an electric field (**E**) and an electric displacement field (**D**), which appear on employing the former on a dielectric material, can be linked by the dielectric constant ε_r.

$$D = \varepsilon_0 \varepsilon_r E \tag{2.23}$$

This dielectric constant demonstrates the ability of a dielectric material to screen the external electric field via the appearance of polarization. This polarization is induced by the reorientation of the electronic density or the motion of the ions forming the material. ε_∞ is noted for the dielectric constant induced by the electronic density and ε_{vib} for the contribution of the dielectric constant involving the ionic movements. Hence,

$$\varepsilon_r = \varepsilon_\infty + \varepsilon_{vib} \tag{2.24}$$

The authors reported using a coupled-perturbed Hartree–Fock/Kohn–Sham (CPKS) scheme adapted to periodic systems to work out the electronic contribution to the static dielectric tensor. This is an unsettling, autonomous technique that is designed to explain the relaxation under the influence of an external electric field. The perturbed wave function is then employed as energy derivatives to calculate the dielectric properties.

The vibrational contribution was calculated according to the next formula [70] from the phonon spectrum. ν_p is the mode p's phonon frequency, V is the unit cell volume, and Z_p is the mass-weighted effective Born vector. For a specific mode, the intensity I_p of IR absorbance, p, is proportional to $|Z_p|^2$.

$$\varepsilon_r = \varepsilon_\infty + \varepsilon_{vib} = \varepsilon_\infty + \frac{4\pi}{V} \sum_p \frac{Z_p^2}{\nu_p^2} \tag{2.25}$$

c. **Effective Mass:** The effective mass can be devised using the band extrema (either at the bottom of the CB or the top of the VB [71]):

$$E(k) = E(k_0) \pm \frac{h^2}{m_\parallel^*}(k_0 - k_\parallel)^2 + \frac{h^2}{m_\perp^*}(k_0 - k_\perp)^2 \tag{2.26}$$

d. **The Exciton's Binding Energy:** When the exciton may be viewed as a hydrogen atom when it is delocalized on multiple unit cells, it is called the

Wannier exciton [72]. Hence, the binding energy of the exciton is the equivalent to the energy of the 1S state of the exciton in the model, and it can be determined using the formula [73]:

$$E_b = E_H \frac{\mu}{\varepsilon_r^2} \qquad (2.27)$$

where E_H is the energy of the hydrogen 1s orbital (13.6 eV) and represents the reduced mass of the exciton.

$$\frac{1}{\mu} = \frac{1}{m_e^*} + \frac{1}{m_h^*} \qquad (2.28)$$

The actual masses of the hole and the electron are measured in multiple crystallographic directions by the average effective mass of the cubic frameworks. The effective masses of the electron and the hole and dielectric constant for non-cubic structures are determined by the equations [73]:

$$\frac{1}{m_e^*} = \frac{2}{3m_{e,\perp}^*} + \frac{\varepsilon_{r,\perp}}{3\varepsilon_{r,\parallel} m_{e,\parallel}^*} \qquad (2.29)$$

$$\frac{1}{m_h^*} = \frac{2}{3m_{h,\perp}^*} + \frac{\varepsilon_{r,\perp}}{3\varepsilon_{r,\parallel} m_{h,\parallel}^*} \qquad (2.30)$$

$$\varepsilon_r = \sqrt{\varepsilon_{r,\perp} \varepsilon_{r,\parallel}} \qquad (2.31)$$

Wannier's model remains no longer valid if the exciton replicates Frenkel's exciton or excitons of molecular crystals, thus having a localized behavior. However, it can be used as a basic first step in estimating the exciton binding energy's magnitude. The Wannier's exciton remains universally valid as the exciton is delocalized on at least three unit cells in the solids considered in this research.

2.3.1.2 Results and Discussions

a. **Cell Parameters:** For the study of SC systems, it is essential to reproduce cells by DFT calculations excellently. The discrepancy between the approaches developed and the experiments done might be inimical to the calculation of other properties, especially in the case of dielectric constant, where cell parameters play an important role.

As the difference between these functionals is very small for this property, only the mean absolute error (MAE) is shown in Figure 2.1. The error was not even 1% for all functionals, except for CdSe and CdTe only, since the maximum error was also less than 3% (gained by B3LYP for CdTe).

PBE0 provided the best results with an MAE of just 0.42%, while HSE6 and PBE also showed acceptable agreements. Surprisingly, where B3LYP is one of the best present functional for chemistry and SC studies, it shows the largest MAE [74–78].

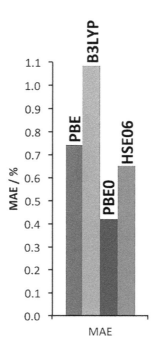

FIGURE 2.1 Mean absolute error on the measured cell parameters (in %). (Reprinted (adapted) with permission from (Bahers, T.L., Rerat, M., Sautet, P., *J. Phys. Chem. C*, 118, 5997–6008, 2014, 1932–7455). Copyright (2014) American Chemical Society.)

b. **Bandgaps:** Bandgap is an essential property for application in a semiconductor. It is worth noting that there is an optimal bandgap between 1.4 and 1.1 eV for a photovoltaic device (for an unpretentious p–n crossing). The bandgap should ideally be between 2.2 and 1.8 eV [79] for the water splitting application. The set of SCs incorporates a broad scope of E_g values between 5.50 eV (C) and 0.66 eV (Ge).

The calculated values represented in Figure 2.2 can be used to extract general trends. The pure GGA functional (PBE) always provides an inferior estimation of the bandgap than that of the experimental value. Global hybrid functions (B3LYP and PBE0) are instead used always to calculate bandgaps higher than empirical references, and greater use of Hartree–Fock exchange within side global hybrid functionals amplifies the bandgap value [80,81]. There is an observable discrepancy that exists between the computed gaps in the CdSe. CdS and CdTe series using global hybrid functionals as the bandgap increases while the GGA functional PBE estimates improve and lower the bandgap to ideal values.

The CdS compound is reasonably well-positioned for the water splitting application with a gap near the optimal area (being between 1.8 and 2.2 eV), which decodes its extensive research for the compound in this application [82–84].

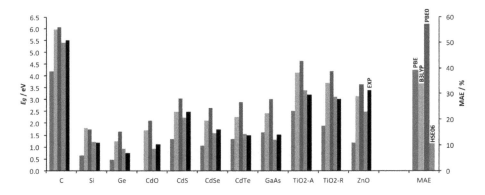

FIGURE 2.2 Calculated gaps (eV) and their MAE (%). (Reprinted (adapted) with permission from (Bahers, T.L., Rerat, M., Sautet, P., *J. Phys. Chem. C*, 118, 5997–6008, 2014, 1932–7455). Copyright (2014) American Chemical Society.)

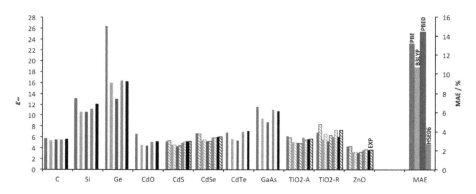

FIGURE 2.3 Calculated ε_∞ with the MAE (in %). (Reprinted (adapted) with permission from (Bahers, T.L., Rerat, M., Sautet, P., *J. Phys. Chem. C*, 118, 5997–6008, 2014, 1932–7455). Copyright (2014) American Chemical Society.)

c. **Dielectric Constants:** The dielectric constant assumes an essential position in the dissociation of excitons as the increase of ε_r diminishes the electrostatic force between the electron and the hole. When an electric field is applied, the dielectric constant is the charges that make up the SC to rearrange the electron and hole (contributed by the exciton).

Figure 2.3 shows ε_∞, which includes both the electronic and vibrational contributions. Figure 2.4 shows the total ε_r, which provides for both the electronic and vibrational contributions.

All the given values of ε_∞ are listed in fair agreement with the experiment from a functional standpoint. As the bandgap that PBE calculates is generally smaller, it provides ε_∞ values greater than other functionals. For example, with an MAE of only 2.7%, E_g calculations using HSE06 outclass the other functionals to compute this property.

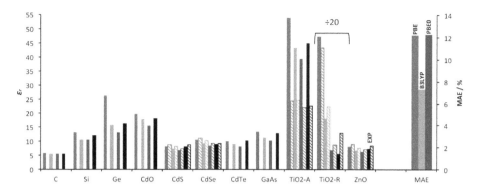

FIGURE 2.4 Calculated ε_r with the MAE (%). ("Reprinted (adapted) with permission from (Bahers, T.L., Rerat, M., Sautet, P., *J. Phys. Chem. C*, 118, 5997–6008, 2014, 1932-7455). Copyright (2014) American Chemical Society.)

d. **Effective Mass:** The effective masses of holes and electrons bear for transport properties. It is connected to CB curvature (for electrons) or VB (for holes) at the extremum of the band.

The lowest computed experimental values of m^* for electrons and holes are shown in Figures 2.5 and 2.6.

Experimental data for effective masses suffer a significant uncertainty with the values shifting from one data series to another frequently. For instance, values can range from 0.12 to 0.35 [85] for light hole effective mass in CdTe. This discrepancy has to do with the difficulty of obtaining effective masses experimentally. Therefore, for this property, the MAE is not calculated between theory and experiment.

The experimentally calculated m^* has the same order of magnitude in comparison with theoretical results. In general, the PBE, B3LYP, and

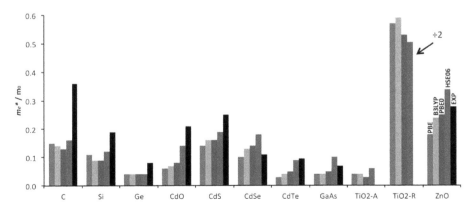

FIGURE 2.5 m_e^* is the smallest value of the computed electron that suffices. (Reprinted (adapted) with permission from (Bahers, T.L., Rerat, M., Sautet, P., *J. Phys. Chem. C*, 118, 5997–6008, 2014, 1932-7455). Copyright (2014) American Chemical Society.)

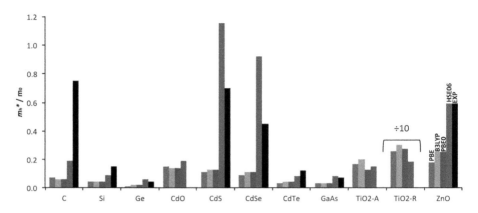

FIGURE 2.6 The computed hole effective masses' smallest value, m_k^*. (Reprinted (adapted) with permission from (Bahers, T.L., Rerat, M., Sautet, P., *J. Phys. Chem. C*, 118, 5997–6008, 2014, 1932–7455). Copyright (2014) American Chemical Society.)

PBE0 functionals give very similar results, whereas m^* calculated with HSE06 is typically higher and better compatible than obtained by the other functionals. The data from DFT also replicate the anisotropies observed experimentally.

e. **Exciton Binding Energy:** The binding energy of the exciton needs to be less than the thermal energy (25 meV at room temperature) to attain a potent separation of this photogenerated exciton. The authors thought that the time of the exciton dissociation is greater than that of the atomic motions for photovoltaic or photochemical devices, and, therefore, E_b corresponds to the binding energy of the relaxed exciton. But doubt remains as these values are prone to influences by external factors such as the temperature and the existence of impurities in order to conclude about E_b's experimental determination.

Both the unrelaxed exciton and relaxed binding energies can be evaluated using ε_∞ and ε_r, respectively. Figure 2.7 represents the E_b values corresponding to the relaxed exciton.

In Figure 2.7, the E_b values obtained with m^* and ε_r computed at the DFT level are collected. Since $\varepsilon_r = \varepsilon_\infty$ E_b is determined by HSE06 for the C, Ge, and Si sequence. For the other species, the experimental value of ε_r needs to be measured to calculate E_b with HSE06. Looking at the excellent agreement between HSE06 and the experimental results, this is a good approximation.

PBE usually produces much low values (sometimes by a factor of 2) than experimental ones for its overestimation of the importance of ε_r. It is, therefore, consistent with the fact that hybrid functions replicate ε_r and m^* reasonably well, and consequently, it is not surprising that B3LYP, PBE0, and HSE06 offer consistent results for the measurement of E_b.

The combination of all the characteristics calculated on TiO_2 anatase and the rutile helps us to better understand the different results observed in the two phases and also helps decipher that anatase is much more efficient than rutile [86–88]. Figure 2.8

Advanced Materials Simulation

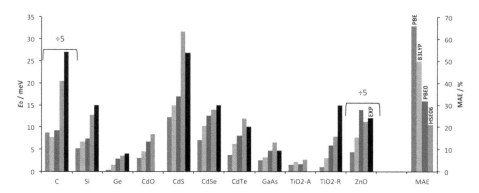

FIGURE 2.7 MAE (%) with exciton binding energy (meV). (Reprinted (adapted) with permission from (Bahers, T.L., Rerat, M., Sautet, P., *J. Phys. Chem. C*, 118, 5997–6008, 2014, 1932–7455). Copyright (2014) American Chemical Society.)

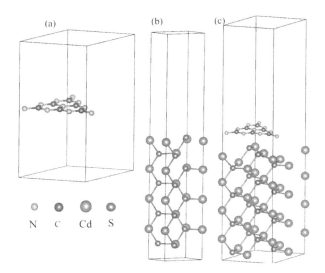

FIGURE 2.8 Geometric structures designed for studying (a) g-C_3N_4 monolayers, (b) CdS (110) surfaces, and (c) g-C_3N_4/CdS heterojunctions. (Reprinted (adapted) with permission from (Liu, J., *J. Phys. Chem. C.*, 119, 28417–28423, 2015). Copyright (2015) American Chemical Society.)

shows Geometric structures designed for studying g-C3N4 monolayers, CdS (110) surfaces, and (c) g-C3N4/CdS heterojunctions.

2.3.2 Investigated Efficiency of Photocatalysis in a Monolayer G-C_3N_4/CdS Heterostructure [89]

Figure 2.9 represents the band structures. The calculated bandgap of 2.76 eV (Figure 2.9a) agrees on both the previously computed value and the experimental

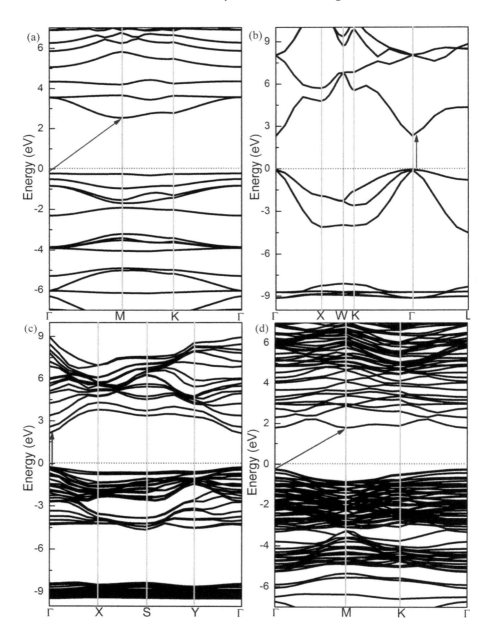

FIGURE 2.9 HSE06 functional used for band structure calculation of (a) monolayer g-C_3N_4, (b) bulk CdS, (c) CdS (110) slab, and (d) g-C_3N_4/CdS heterojunction. ("Reprinted (adapted) with permission from (Liu, J., *J. Phys. Chem. C.*, 119, 28417–28423, 2015). Copyright (2015) American Chemical Society.")

value (2.7 eV) [111]. For bulk CdS, the direct bandgap value of 2.36 eV (Figure 9b) is in fine agreement with experimental data (2.40 eV) [111]. The CdS(110) surface has a direct bandgap of 2.36 eV (Figure 2.9c), and the g-C_3N_4/CdS(110) heterostructure has an indirect gap (2.02 eV) (Figure 2.9d).

Figure 2.10a–c shows the project density of state (PDOS) and total density of state (TDOS) of the g-C_3N_4/CdS heterojunction. The g-C_3N_4/CdS heterojunction can be considered a standard type-II band alignment structure based on experimental studies.

The work function is a critical parameter used to define band alignment [90] and can be defined as:

$$\varphi = E_{vac} - E_F \tag{2.32}$$

Figure 2.11a–c demonstrates the determined electrostatic potentials.

The semiconductor's redox potential is measured by the VBs and CBs' position [91]. The band edge potentials are determined accordingly:

$$E_{VB} = \chi - E_e + 0.5 E_g \tag{2.33}$$

$$E_{CB} = E_{VB} - E_g \tag{2.34}$$

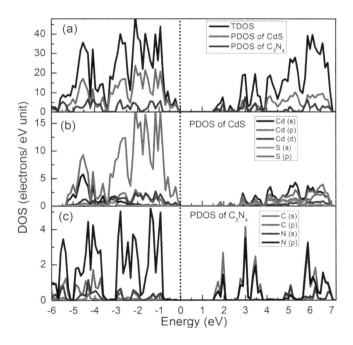

FIGURE 2.10 (a) Using the HSE06 functional, we calculated the TDOS of a g-C_3N_4/CdS heterojunction as well as the PDOS of CdS and g-C_3N_4. (b) In CdS, the measured PDOS of Cd and S atoms. (c) In g-C_3N_4, the measured PDOS of C and N atoms. The Fermi level is the vertical line. (Reprinted (adapted) with permission from (Liu, J., *J. Phys. Chem. C.*, 119, 28417–28423, 2015). Copyright (2015) American Chemical Society.)

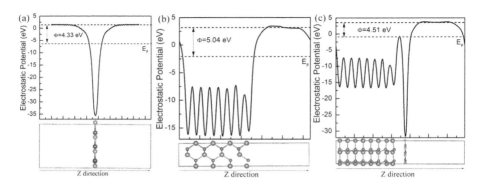

FIGURE 2.11 (a) Monolayer g-C_3N_4, (b) CdS(110) surface, and (c) g-C_3N_4/CdS heterojunction slab model electrostatic potentials. (Reprinted (adapted) with permission from (Liu, J., *J. Phys. Chem. C.*, 119, 28417–28423, 2015). Copyright (2015) American Chemical Society.)

g-C_3N_4 and CdS have χ values 4.73 and 5.18 eV [92,93], respectively. For monolayer g-C_3N_4 and CdS, the measured bandgaps are 2.76 and 2.36 eV, respectively. In contrast to the standard hydrogen electrode (NHE), the above equation indicates that the positions of band edge of CB and VB of g-C_3N_4 are −1.15 and 1.61 V, respectively. CB and VB of CdS have positions of band edge of 0.50 and 1.86 V, respectively, relative to the NHE. The developments are shown in Figure 2.12.

With the Fermi level energy transformations, the VB and CB of g-C_3N_4 relocated downward to 0.17 eV downward, while the CdS's VB and CB relocated upward to 0.53 eV upward. The VB offset (VBO), the relative position of VBM, and the CB offset (CBO), equivalent for the CBM, are pivotal to understand the photocatalytic performance.

After the contact, the edge position of the CB of g-C_3N_4 shifted to −0.98 V, and being lower than H^+/H_2 (0 V), H^+ was reduced to H_2. The edge position of the VB of CdS also shifted to 1.33 V, which was although more than that of O_2/H_2O (1.23 V) was perhaps not apt for the oxidation of water due to overpotential. Hence, the hydrogen evolution reactions had to be promoted using a sacrificial agent in the hydrogen production experiment with g-C_3N_4/CdS heterostructure [94]. The CB edge position of g-C_3N_4 stood at −0.98 V, more negative than that of O_2/O_{2^-} [95] for the g-C_3N_4/CdS heterostructure. Hence, the heterostructure of g-C_3N_4/CdS demonstrated strong photocatalytic efficacy for decomposed organic pollutants under the evolution of photocatalytic hydrogen.

2.3.3 Perceptions from a DFT+U Investigation on Interfacial Interactions, Charge Transport, and Synergic Mechanisms in $BiNbO_4/MWO_4$ (M = Cd and Zn) Heterostructures for Hydrogen Processing [96]

2.3.3.1 Computational Details

Opoku et al. [97,98] conducted DFT calculations [99] on the bulk $ZnWO_4$, $BiNbO_4$, $CdWO_4$, $BiNbO_4/CdWO_4$(010), and $BiNbO_4/ZnWO_4$(010) heterostructures. The

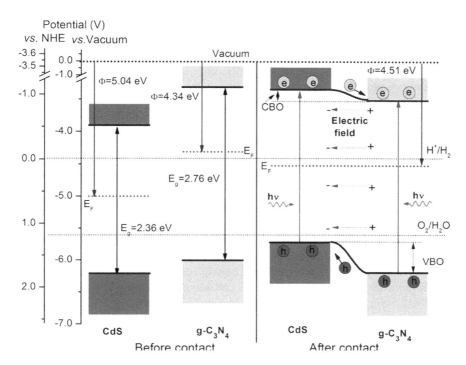

FIGURE 2.12 Positions of the band edges before and after g-C$_3$N$_4$ and CdS contact. (Reprinted (adapted) with permission from (Liu, J., *J. Phys. Chem. C.*, 119, 28417–28423, 2015). Copyright (2015) American Chemical Society.)

ion-electron interaction and exchange-correlation potential are handled by the ultrasoft pseudopotential [100] and the Perdew–Burke–Ernzerhof (PBE) scheme of the comprehensive gradient approximation, respectively. In all model systems, the kinetic energy cutoff was set to 600 eV. The GGA+U approach was used to overcome the incapability of GGA functional to represent systems with localized d and f electrons [101]. The exact bandgap and lattice parameters for bulk BiNbO$_4$, CdWO$_4$, and ZnWO$_4$ were computed using this approach and then be compared to previous experimental data.

2.3.3.2 Results and Discussions

a. **Geometric Structures and Electronic Properties:**
   Before analyzing the electronic structure, a brief discussion needs to be conducted on the interface geometry. The lattice constants value is in agreement with the experimental structural parameters after a little optimization of the bulk ZnWO$_4$ and CdWO$_4$ structures of $a = 4.693(1)$ Å, $b = 5.721(1)$ Å, $c = 4.928(1)$ Å, $V = 132.30$ Å^3, $\beta = 90.63(0)$, and $a = 5.013(7)$ Å, $b = 5.090(10)$ Å, $c = 5.866(9)$ Å, $V = 149.60$ Å, and $\gamma = 91.50°$ (1) [102], respectively. The lattice constants of orthorhombic BiNbO$_4$ also agree well with the available experimental lattice parameters of $a = 5.679$ Å, $b = 11.708$ Å, $c = 4.982$ Å, $V = 331.29$ Å^3, and $\gamma = 90.00°$ [103]. ZnWO$_4$ has a much shorter Zn–O bond

length (2.12 Å) than the Cd–O bond length (2.43 Å) of $CdWO_4$. The Nb–O bond length of 2.02 Å is less than the bond length of Zn (Cd)–O [104]. The bulk $CdWO_4$, $ZnWO_4$, and $BiNbO_4$ bandgaps are then computed using other exchange-correlation functionals. The GGE-PBE functional-determined bandgaps of 2.90 eV for bulk $ZnWO_4$, 2.98 eV for bulk $CdWO_4$, and 3.15 eV for bulk $BiNbO_4$ are underestimated compared to the experimental-established values of 3.80, 3.75, and 3.50 eV [105,106] for bulk $CdWO_4$, $ZnWO_4$, and $BiNbO_4$, respectively. For bulk $CdWO_4$, $ZnWO_4$, and $BiNbO_4$, GGA+U offers valid and realistic bandgap energies of 3.80, 3.75, and 3.48 eV. It matched the experimental values that indicated the correctness and authenticity of the GGA+U procedure and parameters used in this analysis.

The electronic bandgaps tend to expand as the number of on-site Coulomb interactions increase. A fair account of the experimental bandgap is presented by the U_{eff} values of 10.00, 11.00, and 9.00 eV for Cd^{2+}, Zn^{2+}, and Nb^{5+}, correspondingly. Moreover, the 5d-orbital of W is applied a U– value of 3.00 [107]. For Bi, DFT specifies the electronic states for the oxidation states for Bi, and, therefore, +U correction is not necessary here [108]. The adhesion energy, which provides considerable insight into $BiNbO_4$ and MWO_4 interactions, can be measured according to [109]

$$E_{ad} = E_{BiNbO4/MWO4(010)} - E_{BiNbO4} - E_{MWO4} \qquad (2.35)$$

where E_{BiNbO4}, $E_{BiNbO4/MWO4(010)}$, and E_{MWO4} are the total energies of the relaxed, bulk $BiNbO_4$, $BiNbO_4/MWO_4(010)$ heterostructures, and MWO_4, respectively, the interface published as −0.26 and −0.58 eV/Å² for the $BiNbO_4/ZnWO_4(010)$ and $BiNbO_4/CdWO_4(010)$ heterostructures, correspondingly [110,111]. Both processes are exothermic; so, it is energetically favorable to form $BiNbO_4$ on the MWO_4 surface. Furthermore, the $BiNbO_4/ZnWO_4(010)$ heterostructure can form more easily than the $BiNbO_4/CdWO_4(010)$ heterostructure because of lower adhesion energy.

b. **Band Structure:**

Bulk $ZnWO_4$ ($CdWO_4$) has a lattice constant ~18% (~12%) lower than $BiNbO_4$. In the bilayer system, a slight discrepancy in strain value is required to achieve good results [112], particularly because strain may induce a transition from a direct bandgap to an indirect bandgap transition [113] and transition from a semiconductor to a metal [114].

In this investigation, the useful stretched strain decreases the bandgaps energy of the strained monolayers $CdWO_4(010)$, $ZnWO_4(010)$, and $BiNbO_4/MWO_4(010)$ heterostructures. In unstrained systems, the bandgaps of the relaxed $CdWO_4(010)$ surface, $ZnWO_4(010)$ surface, $BiNbO_4/CdWO_4(010)$, and $BiNbO_4/ZnWO_4(010)$ heterostructures are 3.44, 3.25, 2.76, and 2.50 eV, respectively; in strained systems, the bandgaps are 3.23, 2.91, 54, and 2.35 eV, respectively. The direct bandgap values of $ZnWO_4(010)$ and $CdWO_4(010)$ surfaces are 2.91 and 3.23 eV, respectively. The $BiNbO_4(010)$ electrode, on the other hand, is a 3.02 eV indirect band semiconductor. The VBM and CBM of $BiNbO_4/ZnWO_4(010)$ and $BiNbO_4/CdWO_4(010)$ heterostructures have an indirect band difference of 2.35 and 2.54 eV, respectively. As a result, the $BiNbO_4$ and MWO_4-dependent have built-in potential, which is in the range of 0.3–0.5 eV, has a heterostructure value of 0.35–0.4 eV; this heterostructure has $BiNbO_4$ and $CWTbO_4$ in a value of 0.3–0.5 eV with a separation of 0.4 and 0.2 and 0.7 Angstrom to zero degrees of electron overlaps, respectively [115]. These propositions make the devised $BiNbO_4/MWO_4(010)$ heterostructures are a productive approach to designing MWO_4-dependent heterostructures.

c. **Projected Density of State:**

The GGA+U method determined-projected density of states (PDOS) for $BiNbO_4/MWO_4(010)$ heterostructures is used in the analysis of the orbital contributions, electronic coupling, and the effect of a $BiNbO_4$ monolayer on the electronic structure of MWO_4. A comparison is also necessary and has been examined for bulk $ZnWO_4$, $CdWO_4$, and $BiNbO_4$.

Unstrained and strained orbital involvement to VBM and CBM of $ZnWO_4(010)$ surface, $CdWO_4(010)$ surface, $BiNbO_4/ZnWO_4(010)$ heterostructure, $BiNbO_4(010)$ surface, and $BiNbO_4/CdWO_4(010)$ heterostructure are in unison. The PDOS signifies that the CBM is all localized on the $BiNbO_4$ monolayer for both heterostructures. VBM is localized on the MWO_4 monolayers, which can be used for separating charge carriers, light detection, and harvesting since it forms type-II band alignment [116].

d. **Work Function**:

A significant electronic property that determines both Fermi level's relative location and band alignment at the interface is the work function of any photocatalyst material [117]. The work function (Φ) is the benchmark of energy measure needed to extract an electron from the Fermi surface

$$\varphi = E_{vacuum} - E_{vermi} \quad (2.36)$$

where E_{Fermi} and E_{vacuum} are the energy of a stationary electron at the Fermi level and in the vacuum adjacent to the surface, respectively.

The work functions for $ZnWO_4(010)$, $CdWO_4(010)$, $BiNbO_4(010)$, $BiNbO_4/ZnWO_4(010)$, and $BiNbO_4/CdWO_4(010)$, respectively, are calculated to be 7.68, 7.28, 6.20, 5.88, and 6.59 eV in the GGA+U scheme. Therefore, DFT reports suggest that integrating the $BiNbO_4$ monolayer can improve the photocatalytic stability and operation of MWO_4.

e. **Optical Properties:**

The electronic structure of a photocatalyst semiconductor material determines its optical absorption, which is a key factor in measuring photocatalytic performance [118].

When compared to bulk $ZnWO_4$ and $CdWO_4$, $CdWO_4(010)$ heterostructures can be read at 490 and 460 nm, respectively.

In conclusion, coupling $BiNbO_4$ with MWO_4 improves visible light absorption, which is the most significant factor in improving bulk MWO_4 photocatalytic activity.

f. **Possibilities of H_2 Generation:**
Because of their direct bandgap, tunable electronic properties, and outstanding optical absorption, the $BiNbO_4/MWO_4(010)$ heterostructures are sufficiently decent materials for photocatalytic water splitting [119]. In addition, lower adhesion energy is required to promote the manufacture of photocatalyst material. To be considered as an ideal photocatalyst material for water splitting, any semiconductor to enhance solar absorption should have a bandgap <3.00 eV [120]. The main condition is that the VBM band position requirements to be advanced than the reduction potential of O_2/H_2O (0.82 V vs. Standard Hydrogen Electrode (SHE), when pH = 7) and the CBM energy lesser than the H^+/H_2 oxidation potential (−0.41 V vs. SHE) [121]. But the band edges that cannot easily adapt to the H_2O redox potentials for the CB minimum and VB maximum obtained by DFT are not on the absolute scale [122]. GGA+U technique as shown in the following equations [123]:

$$E_{CBM} = E_{BGC} + 0.5 E_g \qquad (2.37)$$

$$E_{VBM} = E_{BGC} - 0.5 E_g \qquad (2.38)$$

where E_{VBM} and E_{CBM} signify the energy of the VBM and CBM, respectively. By changing the vacuum energy scale to zero, the alignments of the CB minimum and the VB maximum energy levels of the different materials are determined. The calculation of band edge for bulk materials is a hard task since it involves the use of thick slabs to indicate the bulk interior and its surface [124]. However, monolayers do not incorporate these shortcomings, terming them ideal systems for band edge calculations.

Since the majority of semiconductors have extremely low exciton binding energies, it is often ignored how optical bandgaps differ from electronics. The driving force is grounded on the potential of O_2 evolution ($E(O_2/H_2O) = 0.82$ V vs. SHE) and E_{VBM} of the semiconductor [125], and so, E_{df} can be determined according to the equation below:

$$E_{df} = (0.82V - (-4.44V))e - E_{VBM} = 5.26 \text{ eV} - E_{VBM} \qquad (2.39)$$

where e denotes an electron charge.

In this analysis, a hybrid $BiNbO_4/MWO_4(010)$ heterostructures were pioneered to verify their photocatalytic activity for H_2 evolution.

2.4 CONCLUSION AND FUTURE PROSPECTS

DFT simulations are excellent tools for calculating material properties theoretically for use in photovoltaics and heterostructures. With more functionals evolving each day, DFT has become more reliable and error-free. The initial functionals prepared had many deviations from the real value, and an error-free estimate was not possible back then. But the latter ones were more complex, as they included a greater number of parameters and were more time-consuming. Theoretical simulations show an essential role in analyzing properties, and thus it is worth pursuing research in this field. Although complexities arise, DFT simulations have the capacity to gather importance, and functionals should be developed involving fewer parameters and more accuracy in solving the required material properties. Photovoltaic systems and heterostructures require a more theoretical basis in development and DFT simulations.

ACKNOWLEDGMENT

This work was supported by the Department of Science and Technology (DST), Central Government of India funded project under the MI IC#5 'Conversion of Sunlight to Storable fuels' issued by DBT-DST Joint Funding Opportunity, Central Government of India through Mission Innovation Programme DST(DST/TMD(EWO)/IC5-2018/06(G)) (S. Roy).

REFERENCES

1. Xong, J., Liu, Y., Liang, S., Zhang, S., Li, Y., Wu, L., Insights into the role of Cu in promoting photocatalytic hydrogen production over ultrathin HNb_3O_8 nanosheets. *J. Catal.*, 342, 98–104, 2016.
2. Brst et al., CdTe solar cells with open-circuit voltage breaking the 1 V barrier. *Nat. Energy*, 1, 16015, 2016.
3. Kwashima, K., Tamai, Y., Ohkita, H., Osaka, I., Takimiya, K., High-efficiency polymer solar cells with small photon energy loss. *Nat. Commun.*, 6, 10085, 2015.
4. Wang, W., Tade, M.O., Shao, Z., Research progress of perovskite materials in photocatalysis- and photovoltaics-related energy conversion and environmental treatment. *Chem. Soc. Rev.*, 44, 5371–5408, 2015.
5. Shah, A., Torres, P., Tscharner, R., Wyrsch, N., Keppner, H., Photovoltaic technology: the case for thin-film solar cells. *Science*, 285, 692–698, 1999.
6. Hagfeldt, A., Boschloo, G., Sun, L., Kloo, L., Pettersson, H., Dye-sensitized solar cells. *Chem. Rev.*, 110, 6595–6663, 2010.
7. Wang, H., Zhang, L., Chen, Z., Junqing, H., Li, S., Wang, Z., Liu, J., Wang, X., *Chem. Soc. Rev.*, 43, 5234–5244, 2014, 1460–4744. Semiconductor heterojunction photocatalysts: design, construction, and photocatalytic performances
8. Fujishima, A., Honda, K., Electrochemical photolysis of water at a semiconductor electrode. *Nature*, 238, 37–38, 1972.
9. Kubacka, A., Fernandez-Garcia, M., Colon, G., Advanced nanoarchitectures for solar photocatalytic applications. *Chem Rev.*, 112, 1555–1614, 2012.
10. Alivisatos, A.P., Semiconductor clusters, nanocrystals, and quantum dots. *Science*, 271, 933–937, 1996.
11. Bahers, T.L., Rerat, M., Sautet, P., Semiconductors used in photovoltaic and photocatalytic devices: assessing fundamental properties from DFT. *J. Phys. Chem. C*, 118, 5997–6008, 2014.
12. Xu, Y., Kraft, M., Xu, R., Metal-free carbonaceous electrocatalysts and photocatalysts for water splitting. *Chem. Soc. Rev.*, 45, 3039–3052, 2016.
13. Cao, S., Yu, J., Carbon-based H_2-production photocatalytic materials. *J. Photochem. Photobiol. C: Photochem. Rev.*, 27, 72–99, 2016.

14. Crabtree, G.W., Dresselhaus, M.S., Buchanan, M.V., The hydrogen economy. *Phys. Today.*, 57, 39–44, 2004.
15. Martha, S., Sahoo, P.C., Parida, K.M., An overview on visible light responsive metal oxide based photocatalysts for hydrogen energy production. *RSC Adv.*, 5, 61535–61553, 2015.
16. Kudo, A., Miseki, Y., Heterogeneous photocatalyst materials for water splitting. *Chem. Soc. Rev.*, 38, 253–278, 2009.
17. Thomas, L.H., The calculation of atomic fields. *Proc. Cambridge Philos. Soc.*, 23, 542–548, 1927.
18. Fermi, E. Z., Eine statistische Methode zur Bestimmung einiger Eigenschaften des Atoms und ihre Anwendung auf die Theorie des periodischen Systems der Elemente. *Phys.*, 48, 73-79, 1928.
19. Slater, J.C., A simplification of the hartree-fock method. *Phys. Rev.*, 81, 385, 1951.
20. Dirac, P.A.M., Note on exchange phenomena in the Thomas atom. *Proc. Cambridge Philos. Soc.*, 26, 376–385, 1930.
21. Ceperley, D.M., Alder, B.J., Ground state of the electron gas by a stochastic method. *Phys. Rev. Lett.*, 45, 566, 1980.
22. Vosko, S.H., Wilk, L., Nusair, M., *Can. J. Phys.*, 45, 1200, 1980, 0008–4204. Accurate spin-dependent electron liquid correlation energies for local spin density calculations: a critical analysis
23. Perdew, J.P., Wang, Y., Accurate and simple analytic representation of the electron-gas correlation energy. *Phys. Rev B.*, 45, 13244, 1992.
24. Gross, E.K.U., Dreizler, R.M.Z., Gradient expansion of the Coulomb exchange energy. *Phys. A: Hadrons Nucl.*, 302, 103–106, 1981.
25. Levy, M., Perdew, J.P., Hellmann-Feynman, virial, and scaling requisites for the exact universal density functionals. Shape of the correlation potential and diamagnetic susceptibility for atoms. *Phys. Rev. A*, 32, 2010, 1985.
26. Levy, M., *Dens. Function. Theory*, 337, 11–31, 1995, 0258–1221.
27. Lee, C., Yang, W., Parr, R.G., Development of the Colle-Salvetti correlation-energy formula into a functional of the electron density. *Phys. Rev. B*, 37, 785, 1988.
28. Perdew, J.P., Schimdt, K., *Density Functional Theory and Its Application to Materials*, VanDoren, V., VanAlsenoy, C., Geerlings, P., Eds., 577, 1–20, 2001.
29. Becke, A.D., Correlation energy of an inhomogeneous electron gas: A coordinate-space model. *J. Chem. Phys.*, 88, 1053, 1988.
30. Becke, A.D., Roussel, M.R., Exchange holes in inhomogeneous systems: A coordinate-space model. *Phys. Rev. A*, 39, 3761, 1989.
31. Tao, J., Perdew, J.P., Staroverov, V.N., Scuseria, G.E., Climbing the density functional ladder: nonempirical meta–generalized gradient approximation designed for molecules and solids. *Phys. Rev. Lett.*, 91, 146401, 2003.
32. Becke, A.D., A new mixing of Hartree–Fock and local density-functional theories. *J. Chem. Phys.*, 98, 1372, 1993.
33. Stephens, P.J., Devlin, F.J., Chabalowski, C.F., Frisch, M.J., Ab Initio calculation of vibrational absorption and circular dichroism spectra using density functional force fields. *J. Phys. Chem.*, 98, 11623–11627, 1994.
34. Savin, A., Flad, H.J., Density functionals for the Yukawa electron-electron interaction. *Int. J. Quantum Chem.*, 56, 327–332, 1995.
35. Leninger, T., Stoll, H., Werner, H.J., Savin, A., Combining long-range configuration interaction with short-range density functionals. *Chem. Phys. Lett.*, 275, 151–160, 1997.
36. Toulouse, J., Colonna, F., Savin, A., Long-range–short-range separation of the electron-electron interaction in density-functional theory. *Phys. Rev. A*, 70, 062505, 2004.
37. Gill, P.M.W., Adamson, R.D., A family of attenuated Coulomb operators. *Mol. Phys.*, 261, 105–110, 1996.

38. Adamson, R.D., Dombrowski, J.P., Gill, P.M.W., Chemistry without Coulomb tails. *Chem. Phys. Lett.*, 254, 329–336, 1996.
39. Iikhura, H., Tsuneda, T., Yanai, T., Hirao, K., A long-range correction scheme for generalized-gradient-approximation exchange functionals. *J. Chem. Phys.*, 115, 3540, 2001.
40. Morrell, M.M., Parr, R.G., Levy, M., Calculation of ionization potentials from density matrices and natural functions, and the long-range behavior of natural orbitals and electron density. *J. Chem. Phys.*, 62, 549, 1975.
41. Peach, M.J.G., Benfield, P., Helgaker, T., Tozer, D.J., Excitation energies in density functional theory: An evaluation and a diagnostic test. *J. Chem. Phys.*, 128, 044118, 2008.
42. Becke, A.D., Density-functional thermochemistry. V. Systematic optimization of exchange-correlation functionals. *J. Chem. Phys.*, 107, 8554, 1997.
43. Schmider, H.L., Becke, A.D., Optimized density functionals from the extended G2 test set. *J. Chem. Phys.*, 108, 9624, 1998.
44. Chan, G.K.L., Handy, N.C., An extensive study of gradient approximations to the exchange-correlation and kinetic energy functionals. *J. Chem. Phys.*, 112, 5639, 2000.
45. Becke, A.D., A new inhomogeneity parameter in density-functional theory. *J. Chem. Phys.*, 109, 2092, 1998.
46. Boese, A.D., Handy, N.C., New exchange-correlation density functionals: The role of the kinetic-energy density. *J. Chem. Phys.*, 116, 9559, 2002.
47. Zhao, Y., Truhlar, D.G., A new local density functional for main-group thermochemistry, transition metal bonding, thermochemical kinetics, and noncovalent interactions. *J. Chem. Phys.*, 125, 194101, 2006.
48. Zhao, Y., Truhlar, D.G., Density functionals with broad applicability in chemistry. *Acc. Chem. Res.*, 41, 157–167, 2008.
49. Langreth, D.C., Perdew, J.P., Exchange-correlation energy of a metallic surface: Wavevector analysis. *Phys. Rev. B*, 15, 2884, 1977.
50. Seidl, M., Perdew, J.P., Kurth, S., Density functionals for the strong-interaction limit. *Phys. Rev. A*, 62, 012502, 2000.
51. Perdew, J.P., Kurth, S., Seidl, M., Exploring the adiabatic connection between weak- and strong-interaction limits in density functional theory. *Int. J. Mod. Phys. B*, 15, 1672–1683, 2001.
52. Ernzerhof, M., Perdew, J.P., Burke, K., Coupling-constant dependence of atomization energies. *Int. J. Quantum Chem.*, 64, 285-295, 1997.
53. Mori-Sanchez, P., Cohen, A.J., Yang, W.T., Self-interaction-free exchange-correlation functional for thermochemistry and kinetics. *J. Chem. Phys.*, 124, 091102, 2006.
54. Teale, A.M., Coriani, S., Helgaker, T., Range-dependent adiabatic connections. *J. Chem. Phys.*, 133, 164112, 2010.
55. Teale, A.M., Coriani, S., Helgaker, T., Accurate calculation and modeling of the adiabatic connection in density functional theory. *J. Chem Phys.*, 132, 164115, 2010.
56. Colonna, F., Savin, A., Correlation energies for some two- and four-electron systems along the adiabatic connection in density functional theory. *J. Chem. Phys.*, 110, 2828, 1999.
57. Savin, A., Colonna, F., Allavena, M., Analysis of the linear response function along the adiabatic connection from the Kohn–Sham to the correlated system. *J. Chem. Phys.*, 115, 6827, 2001.
58. Wu, Q., Yang, W.T., A direct optimization method for calculating density functionals and exchange–correlation potentials from electron densities. *J. Chem. Phys.*, 118, 2498, 2003.
59. Tao, J.M., Staroverov, V.N., Scuseria, G.E., Perdew, J.P., Exact-exchange energy density in the gauge of a semilocal density-functional approximation. *Phys. Rev. A*, 77, 012509, 2008.
60. Gorling, A., Levy, M., Exact Kohn-Sham scheme based on perturbation theory. *Phys. Rev. A*, 50, 196, 1994.

61. Ren, X., Tkatchenko, A., Rinke, P., Scheffler, M., Beyond the random-phase approximation for the electron correlation energy: The importance of single excitations. *Phys. Rev. Lett.*, 106, 153003, 2011.
62. Mori-Sanchez, P., Wu, Q., Yang, W.T., Orbital-dependent correlation energy in density-functional theory based on a second-order perturbation approach: Success and failure. *J. Chem. Phys.*, 123, 062204, 2005.
63. Grimme, S., Semiempirical hybrid density functional with perturbative second-order correlation. *J. Chem. Phys.*, 124, 034108, 2006.
64. Scuseria, G.E., Henderson, T.M., Sorensen, D.C., The ground state correlation energy of the random phase approximation from a ring coupled cluster doubles approach. *J. Chem. Phys.*, 129, 231101, 2008.
65. Toulouse, J., Gerber, I.C., Jansen, G., Savin, A., Angyan, J.G., Adiabatic-connection fluctuation-dissipation density-functional theory based on range separation. *Phys. Rev. Lett.*, 102, 096404, 2009.
66. Janesko, B.G., Henderson, T.M., Scuseria, G.E., Long-range-corrected hybrid density functionals including random phase approximation correlation: Application to noncovalent interactions. *J. Chem. Phys.*, 133, 179901, 2010.
67. Paier, J., Janesko, B.G., Henderson, T.M., Scuseria, G.E., Gruneis, A., Kresse, G.J., Hybrid functionals including random phase approximation correlation and second-order screened exchange. *Chem. Phys.*, 133, 179902, 2010.
68. Gruneis, A., Marsman, M., Harl, J., Schimka, L., Kresse, G., Making the random phase approximation to electronic correlation accurate. *J. Chem. Phys.*, 131, 154115, 2009.
69. Yang, Z., Ullrich, C., Direct calculation of exciton binding energies with time-dependent density-functional theory. *Phys. Rev. B*, 87, 195204, 2013.
70. Gonze, X., Lee, C., Dynamical matrices, Born effective charges, dielectric permittivity tensors, and interatomic force constants from density-functional perturbation theory. *Phys. Rev. B*, 55, 10355, 1997.
71. Madelung, O., *Semiconductors: Data Handbook*, 3rd ed., Springer, New York, 2004.
72. Pelant, I., Valenta, J., *Luminescence Spectroscopy of Semiconductors*, Oxford University Press, United Kingdom, 2012.
73. Rodina, A., Dietrich, M., Goldner, A., Eckey, L., Hoffmann, A., Efros, A., Rosen, M., Meyer, B., Free excitons in wurtzite GaN. *Phys. Rev. B*, 64, 115204, 2001.
74. Xiao, H., Tahir-Kheli, J., Goddard, W., Accurate band gaps for semiconductors from density functional theory. *J. Phys. Chem. Lett.*, 2, 212–217, 2011.
75. Graciani, J., Marquez, A.M., Plata, J.J., Ortega, Y., Hernandez, N.O., Meyer, A., Zicovich-Wilson, C.M., Sanz, J.F., Comparative study on the performance of hybrid DFT functionals in highly correlated oxides: The case of CeO_2 and Ce_2O_3. *J. Chem. Theor. Comput.*, 7, 56–65, 2011.
76. Zicovich-Wilson, C.M., San-Roman, M.L., Camblor, M.A., Pascale, F., Durand-Niconoff, J.S., Structure, vibrational analysis, and insights into host−guest interactions in As-synthesized pure silica ITQ-12 Zeolite by Periodic B3LYP calculations. *J. Am. Chem. Soc.*, 129, 11512–11523, 2007.
77. Martsinovich, N., Ambrosio, F., Troisi, A., Adsorption and electron injection of the N3 metal–organic dye on the TiO_2 rutile (110) surface. *Phys. Chem. Chem. Phys.*, 14, 16668–16675, 2012.
78. Nilsing, M., Persson, P., Lunell, S., Ojamae, L., Dye-sensitization of the TiO_2 Rutile (110) surface by perylene dyes: Quantum-chemical periodic B3LYP computations. *J. Phys. Chem. C*, 111, 12116–12123, 2007.
79. Moriya, A., Takata, T., Domen, K., Recent progress in the development of (oxy)nitride photocatalysts for water splitting under visible-light irradiation. *Coord. Chem. Rev.*, 257, 1957–1969, 2013.

80. Labat, F., Ciofini, I., Adamo, C., Modeling ZnO phases using a periodic approach: From bulk to surface and beyond. *J. Chem. Phys.*, 131, 044708, 2009.
81. Dahan, N., Jehl, Z., Hildebrandt, T., Greffet, J-J., Guillemoles, J-F., Lincot, D., Naghavi, N., Optical approaches to improve the photocurrent generation in Cu(In,Ga)Se$_2$ solar cells with absorber thicknesses down to 0.5 µm. *J. Appl. Phys.*, 112, 094902, 2012.
82. Osterloh, F.E., Inorganic materials as catalysts for photochemical splitting of water. *Chem. Mater.*, 20, 35–54, 2008.
83. Hernandez-Alonso, M.D., Fresno, F., Suarez, S., Coronado, J.M., Development of alternative photocatalysts to TiO$_2$: Challenges and opportunities. *Energ. Environ. Sci*, 2, 1231–1257, 2009.
84. Li, Q., Guo, B., Yu, J., Ran, J., Zhang, B., Yan, H., Gong, J.R., Highly efficient visible-light-driven photocatalytic hydrogen production of CdS-cluster-decorated graphene nanosheets. *J. Am. Chem. Soc.*, 133, 10878–10884, 2011.
85. Kissling, G.P., Fermin, D.J., Electrochemical hole injection into the valence band of thiol stabilised CdTe quantum dots. *Phys. Chem. Chem. Phys*, 11, 10080–10086, 2009.
86. Liu, L., Zhao, H., Andino, J.M., Li, Y., Photocatalytic CO$_2$ reduction with H$_2$O on TiO$_2$ nanocrystals: Comparison of anatase, rutile, and brookite polymorphs and exploration of surface chemistry. *ACS. Catal.*, 2, 1817–1828, 2012.
87. Magne, C., Dufour, F., Labat, F., Lancel, G., Durupthy, O., Cassaignon, S., Pauporte, T., Effects of TiO$_2$ nanoparticle polymorphism on dye-sensitized solar cell photovoltaic properties. *J. Photochem. Photobiol. A.*, 232, 22–31, 2012.
88. Sclafani, A., Hermann, J.M., Comparison of the photoelectronic and photocatalytic activities of various anatase and rutile forms of titania in pure liquid organic phases and in aqueous solutions. *J. Phys. Chem.*, 3654, 13655–13661, 1996.
89. Liu, J., Origin of high photocatalytic efficiency in monolayer g-C$_3$N$_4$/CdS heterostructure: A hybrid DFT study. *J. Phys. Chem. C*, 119, 28417–28423, 2015.
90. Liu, J., Hua, E., Electronic structure and absolute band edge position of tetragonal AgInS$_2$ photocatalyst: A hybrid density functional study. *Mater. Sci. Semicond. Process.*, 40, 446–452, 2015.
91. Liu, J.J., Fu, X.L., Chen, S.F., Zhu, Y.F., Electronic structure and optical properties of Ag$_3$PO$_4$ photocatalyst calculated by hybrid density functional method. *Appl. Phys. Lett.*, 99, 191903, 2011.
92. Chen, S., Hu, Y., Meng, S., Fu, X., Study on the separation mechanisms of photogenerated electrons and holes for composite photocatalysts g-C$_3$N$_4$-WO$_3$. *Appl. Catal. B*, 150–151, 564–573, 2014.
93. Xu, Y., Schoonen, M.A., The absolute energy positions of conduction and valence bands of selected semiconducting minerals. *Am. Mineral*, 85, 543–556, 2000.
94. Cao, D. et al., In-situ growth of CdS quantum dots on g-C$_3$N$_4$ nanosheets for highly efficient photocatalytic hydrogen generation under visible light irradiation. *Int. J. Hydrogen Energy*, 38, 1258–1266, 2013.
95. Berger, T., Monllor-Satoca, D., Jankulovska, M., Lana-Villarreal, T., Gomez, R., The electrochemistry of nanostructured titanium dioxide electrodes. *ChemPhysChem.*, 13, 2824–2875, 2012.
96. Opoku, F., Govender, K.K., van Sittert, C., Govender, P.P., Charge transport, interfacial interactions and synergistic mechanisms in BiNbO$_4$/MWO$_4$ (M = Zn and Cd) heterostructures for hydrogen production: insights from a DFT+U study. *Phys. Chem. Chem. Phys.*, 19, 28401–28413, 2017.
97. Clark, S.J., Segall, M.D., Pickard, C.J., Hansip, P.J., Probert, M.I., Refson, K., Payne, M.C., First principles methods using CASTEP. *Z. Kristallogr. Cryst. Mater.*, 220, 567–570, 2005.
98. Materials Studio Simulation environment, Release 2016.

99. Kohn, W., Sham, L.J., Self-consistent equations including exchange and correlation effects. *Phys. Rev.*, 140, A1133–A1138, 1965.
100. Vanderbilt, D., Soft self-consistent pseudopotentials in a generalized eigenvalue formalism. *Phys. Rev. B.*, 41, 7892–7895, 1990.
101. Anisimov, V., Korotin, M., Zaanen, J., Andersen, O., Spin bags, polarons, and impurity potentials in $La_{2-x}Sr_xCuO_4$ from first principles. *Phys. Rev. Lett.*, 68, 345–348, 1992.
102. Dahlborg, M.A., Svensson, G., Structural Changes in the System $Zn_{1-x}Cd_xWO_4$, determined from single crystal data. *Acta Chem. Scand.*, 53, 1103–1109, 1999.
103. Sales, A., Oliveira, P., Almeida, J., Costa, M., Rodrigues, H., Sombra, A., Copper concentration effect in the dielectric properties of $BiNbO_4$ for RF applications. *J. Alloys. Compd.*, 542, 264–270, 2012.
104. Shannon, R., Revised effective ionic radii and systematic studies of interatomic distances in halides and chalcogenides. *Acta. Crystalogr. Sect. A.*, 32, 751–767, 1976.
105. Bonanni, M., Spanhel, L., Lerch, M., Fuglein, E., uller, G., Jermann, F., Conversion of colloidal $ZnO-WO_3$ heteroaggregates into strongly blue luminescing $ZnWO_4$ Xerogels and films. *Chem Mater.*, 10, 304–310, 1998.
106. Zou, Z., Arakawa, H., Ye, J., Substitution effect of Ta^{5+} by Nb^{5+} on photocatalytic, photophysical, and structural properties of $BiTa_{1-x}Nb_xO_4 (0 \leqq x \leqq 1.0)$. *J. Mater. Res.*, 17, 1446–1454, 2002.
107. Kamisaka, H., Suenaga, T., Nakamura, H., Yamashita, K., DFT-based theoretical calculations of Nb- and W-doped Anatase TiO_2: Complex formation between W dopants and oxygen vacancies. *J. Phys. Chem. C.*, 114, 12777–12783, 2010.
108. Lucid, A., Iwaszuk, A., Nolan, M., A first principles investigation of Bi_2O_3-modified TiO_2 for visible light Activated photocatalysis: The role of TiO_2 crystal form and the Bi^{3+} stereochemical lone pair. *Mater. Sci. Semicond. Process.*, 25, 59–67, 2014.
109. Du, A., Sanvito, S., Li, D., Wang, D., Jiao, Y., Liao, T., Sun, Q., Ng, Y.H., Zhu, Z., Amal, R., Hybrid graphene and graphitic carbon nitride nanocomposite: Gap opening, electron–hole puddle, interfacial charge transfer, and enhanced visible light response. *J. Am. Chem. Soc.*, 134, 4393–4397, 2012.
110. Yang, Y.C., Xu, L., Huang, W.Q., Luo, C.Y., Huang, G.F., Peng, P., Electronic structures and photocatalytic responses of $SrTiO_3$(100) surface interfaced with graphene, reduced graphene oxide, and graphane: Surface termination effect. *J. Phys. Chem. C.*, 119, 19095–19104, 2015.
111. Xu, L., Huang, W.Q., Wang, L.L., Huang, G.F., Peng, P., Mechanism of superior visible-light photocatalytic activity and stability of hybrid Ag_3PO_4/Graphene nanocomposite. *J. Phys. Chem. C.*, 118, 12972–12979, 2014.
112. Johari, P., Shenoy, V.B., Tuning the electronic properties of semiconducting transition metal dichalcogenides by applying mechanical strains. *ACS Nano*, 6, 5449–5456, 2012.
113. Yun, W.S., Han, S., Hong, S.C., Kim, I.G., Lee, J., Thickness and strain effects on electronic structures of transition metal dichalcogenides: 2H-MX_2 semiconductors (M=Mo, W;X= S, Se, Te). *Phys. Rev. B*, 85, 033305, 2012.
114. Scalise, E., Houssa, M., Pourtois, G., Afanas'ev, V., Stesmans, A., Strain-induced semiconductor to metal transition in the two-dimensional honeycomb structure of MoS_2. *Nano. Res.*, 5, 43–48, 2012.
115. Niu, M., Cheng, D., Cao, D., Understanding the mechanism of photocatalysis enhancements in the graphene-like semiconductor sheet/TiO_2 composites. *J. Phys. Chem. C*, 118, 5954–5960, 2014.
116. Komsa, H.-P., Krasheninnikov, A.V., Electronic structures and optical properties of realistic transition metal dichalcogenide heterostructures from first principles. *Phys. Rev. B*, 88, 085318, 2013.

117. Peng, X., Tang, F., Copple, A., Engineering the work function of armchair graphene nanoribbons using strain and functional species: a first principles study. *J. Phys. Condens. Matter.*, 24, 075501, 2012.
118. Sun, L., Qi, Y., Jia, C.-J., Jin, Z., Fan, W., Enhanced visible-light photocatalytic activity of g-C_3N_4/Zn_2GeO_4 heterojunctions with effective interfaces based on band match. *Nanoscale*, 6, 2649–2659, 2014.
119. Singh, A.K., Mathew, K., Zhuang, H.L., Hennig, R.G., Computational screening of 2D materials for photocatalysis. *J. Phys. Chem. Lett.*, 6, 1087–1098, 2015.
120. Jiao, Y., Zhou, L., Ma, F., Gao, G., Kou, L., Bell, J., Sanvito, S., Du, A., Predicting single-layer technetium dichalcogenides (TcX2, X = S, Se) with promising applications in photovoltaics and photocatalysis. *ACS Appl. Mater. Interfaces.*, 8, 5385–5392, 2016.
121. Kanan, D.K., Carter, E.A., Band gap engineering of MnO via ZnO alloying: A potential new visible-light photocatalyst. *J. Phys. Chem. C*, 116, 9876–9887, 2012.
122. Modak, B., Ghosh, S.K., Improving $KNbO_3$ photocatalytic activity under visible light. *RSC Adv.*, 6, 9958–9966, 2016.
123. Zhuang, H.L., Hennig, R.G., Single-layer Group-III monochalcogenide photocatalysts for water splitting. *Chem. Mater.*, 25, 3232–3238, 2013.
124. Rasmussen, F.A., Thygesen, K.S., Computational 2D materials database: Electronic structure of transition-metal dichalcogenides and oxides. *J. Phys. Chem. C.*, 119, 13169–13183, 2015.
125. Xu, S.M., Pan, T., Dou, Y.B., Yan, H., Zhang, S.T., Ning, F.Y., Shi, W.Y., Wei, M., Theoretical and experimental study on $M^{II}M^{III}$-layered double hydroxides as efficient photocatalysts toward oxygen evolution from water. *J. Phys. Chem. C*, 119, 18823–18834, 2015.
126. Green, M.A., Emery, K., Hishikawa, Y., Warta, W., Dunlop, E.D., Solar cell efficiency tables (version 42). *Prog. Photovolt Res. Appl.*, 21, 827–837, 2013.
127. King, R.R., Law, D.C., Edmondson, K.M., Fetzer, C.M., Kinsey, G.S., Yoon, H., Sherif, R.A., Karam, N.H., 40% efficient metamorphic GaInP/GaInAs/Ge multijunction solar cells. *App. Phys. Lett.*, 90, 183516, 2007.
128. Bosi, M., Pelosi, C., The potential of III-V semiconductors as terrestrial photovoltaic devices. *Prog. Photovolt Res. Appl.*, 15, 51–68, 2007.
129. Wu, X., High-efficiency polycrystalline CdTe thin-film solar cells. *Sol. Energy.*, 77, 803–814, 2004.
130. Jackson, P., Hariskos, D., Lotter, E., Paetel, S., Wuerz, R., Menner, R., Wischmann, W., Powalla, M., New world record efficiency for Cu(In,Ga)Se_2 thin-film solar cells beyond 20%. *Prog. Photovolt Res. Appl.*, 19, 894–897, 2011.
131. Singh, U.P., Patra, S.P., Progress in polycrystalline thin-film Cu(In,Ga)Se_2 solar cells. *Int. J. Photoenergy*, 2010, 1–19, 2010.
132. Guerin, V.M., Magne, C., Pauporte, Th., Bahers, T.L., Rathousky, J., Electrodeposited nanoporous versus nanoparticulate ZnO films of similar roughness for dye-sensitized solar cell applications. *ACS Appl. Mater. Interfaces*, 2, 3677–3685, 2010.
133. Ramasamy, E., Lee, J., Ordered mesoporous SnO_2–based photoanodes for high-performance dye-sensitized solar cells. *J. Phys. Chem. C*, 114, 22032–22037, 2010.
134. Odobel, F., Pleux, L.L., Pellegrin, Y., Blart, E., New photovoltaic devices based on the sensitization of p-type semiconductors: challenges and opportunities. *Acc. Chem. Res.*, 43, 1063–1071, 2010.
135. Kamat, P.V., Quantum dot solar cells. Semiconductor nanocrystals as light harvesters. *J. Phys. Chem. C*, 112, 18737–15753, 2008.
136. Guo, J., Ouyang, S., Kako, T., Ye, Mesoporous In(OH)$_3$ for photoreduction of CO_2 into renewable hydrocarbon fuels. J., *App. Surf. Sci.*, 280, 418-423, 2013.

137. Woolerton, T.W., Sheard, S., Reisner, E., Pierce, E., Ragsdale, S.W., Armstrong, F.A., Efficient and clean photoreduction of CO_2 to CO by enzyme-modified TiO_2 nanoparticles using visible light. *J. Am. Chem. Soc.*, 132, 2132–2133, 2010.
138. Liu, L., Zhao, H., Andino, J.M., Li, Y., Photocatalytic CO2 reduction with H2O on TiO2 nanocrystals: Comparison of anatase, rutile, and brookite polymorphs and exploration of surface chemistry. *ACS Catal.*, 2, 1817–1828, 2012.
139. Liu, L., Gao, F., Zhao, H., Li, Y., Tailoring Cu valence and oxygen vacancy in Cu/TiO_2 catalysts for enhanced CO2 photoreduction efficiency. *Appl. Catal. B-Environ*, 134–135, 349–358, 2013.
140. Sivula, K., Formal, F.L., Gratzel, M., Solar water splitting: Progress using hematite (α-Fe_2O_3) photoelectrodes'. *ChemSusChem*, 4, 432–449, 2011.
141. Li, Q., Guo, B., Yu, J., Ran, J., Zhang, B., Yan, H., Gong, J.R., Highly efficient visible-light-driven photocatalytic hydrogen production of CdS-cluster-decorated graphene nanosheets. *J. Am. Chem. Soc.*, 28, 10878–10884, 2011.
142. Ji, Z., He, M., Huang, Z., Ozkan, U., Wu, Y., Photostable p-Type dye-sensitized photoelectrochemical cells for water reduction. *J. Am. Chem. Soc.*, 135, 11696–11699, 2013.
143. Moriya, Y., Takata, T., Domen, K., Recent progress in the development of (oxy)nitride photocatalysts for water splitting under visible-light irradiation. *Coord. Chem. Rev.*, 257, 1957–1969, 2013.
144. Shang, M., Wang, W., Sun, S., Ren, J., Zhou, L., Zhang, L., Efficient visible light-induced photocatalytic degradation of contaminant by spindle-like PANI/$BiVO_4$. *J. Phys. Chem. C*, 113, 20228–20233, 2009.
145. Konstantinou, I.K., Albanis, T.A., TiO_2-assisted photocatalytic degradation of azo dyes in aqueous solution: kinetic and mechanistic investigations: A review. *Appl. Catal. B-Environ.*, 49, 1–14, 2004.

3 Molecular Dynamics Simulations for Structural Characterization and Property Prediction of Materials

S. K. Joshi
University of Petroleum and Energy Studies

CONTENTS

3.1 Introduction 50
    3.1.1 Force Fields in Molecular Dynamics...51
    3.1.2 Molecular Dynamics Algorithm...51
    3.1.3 Parameters for MD Simulations 52
        3.1.3.1 Starting Structure..52
        3.1.3.2 Initial Temperature/Energy...52
        3.1.3.3 Size of the Timestep..53
        3.1.3.4 Length of a Simulation 53
3.2 Material Simulation at the Nanoscale and Bulk and Its Visualization...........54
    3.2.1 *Atomsk* to Make New Material Structures 54
    3.2.2 LAMMPS Molecular Dynamics Simulator 55
        3.2.2.1 Initialization..55
        3.2.2.2 Atom Definition 55
        3.2.2.3 Settings...56
        3.2.2.4 Running a Simulation 56
        3.2.2.5 Ensembles, Constraints, and Boundary Conditions.............56
3.3 Property Prediction with Molecular Dynamics..57
    3.3.1 Mechanical Properties 57
    3.3.2 Thermal Properties 58
3.4 Post-Processing and the Visualization of the Output Data............................60
    3.4.1 Open Visualization Tool (OVITO) 61
    3.4.2 Visual Molecular Dynamics (VMD)...62

DOI: 10.1201/9781003121954-3

3.5   Machine Learning for Performance Enhancement
      of Molecular Dynamics Simulations 62
      3.5.1   Methodology of Applying Machine Learning in Molecular
              Dynamics Simulations 64
      3.5.2   Machine Learning and Related Approaches
              to Enhance Sampling 64
3.6   Conclusion 65
References 66

3.1   INTRODUCTION

Molecular dynamics (MD) simulation is one of a growing number of methods to study the macroscopic behavior of systems by following the evolution of the system at the molecular scale. It was introduced by Alder and Wainwright [1] in the 1950s to study the interaction of hard spheres, which interact with each other through perfect collisions. However, the credit to actually perform the molecular simulation for the first time goes to Rahman [2] for liquid argon.

MD models properties of a system of interacting particles by repeatedly calculating the interactions between the particles and integrating their equations of motion. This process traces out a discrete phase-space trajectory. Combining statistical mechanics and kinetic theory, microscopic properties of the system can be calculated. Sampling and averaging these properties of the system along a sufficiently long trajectory approximate the corresponding macroscopic properties.

In the MD method, we explore the macroscopic properties of a system through microscopic simulations. At a certain temperature T, atomic movements and initial velocities are assigned as per the Maxwell distribution. The number of particles, timestep, total time, and simulation size duration must be selected so that the calculation can be done within a reasonable time period. Thermodynamic parameters of the system like pressure, volume, temperature, and total energy change as the simulation proceeds. Simulation can be done in various ways, i.e., considering different types of ensembles. A statistical ensemble consists of a collection of all possible states of a system compatible with some set of constraints, such as imposed temperature or imposed number of molecules. Depending on the application desired, different statistical ensembles are used (Table 3.1). Each statistical ensemble is characterized by its probability density, i.e., the probability of occurrence of each system state in the collection.

TABLE 3.1
Statistical Ensembles for Molecular Dynamics Simulations

Statistical Ensemble	Imposed Variables	Applications
Canonical ensemble	N, V, T	Determination of phase properties
Grand canonical ensemble	M, V, T	Adsorption isotherms, selectivities
Isothermic-isobaric ensemble	N, P, T	Phase properties
Micro canonical ensemble	E, V, T	Transport properties

Presently, MD is being used for simulating various cooperative phenomena (friction, wear, indentation, and so on), dynamic properties (like adsorption, diffusion, relaxation, and so on), static properties (defects, interfaces, and so on), thermodynamic properties (entropy, enthalpy, specific heat, temperature, pressure, KE, PE, phase changes, and so on), and structures of bulk materials, solutions, aggregates, DNAs, proteins, clusters, molecules, and so on [3–20]. It can also be used for studying the effects of temperature, pressure, solvents, and intermolecular interactions on a system of particles.

3.1.1 Force Fields in Molecular Dynamics

Theoretically, a molecule is considered a collection of atoms held together by simple elastic or harmonic forces. A force field is a mathematical expression that describes the dependence of the energy of a molecule on the coordinates of the atoms in the molecule. Force fields use simplified functions (potential functions) to describe the interactions between atoms. Force fields are constructed by parameterizing the potential functions using either experimental data (X-ray and electron diffraction, NMR, and IR spectroscopy) or ab initio and semi-empirical quantum mechanical calculations.

It can be considered that the molecular mechanics energy, (E_M), is made up of a number of components. The bond energy (E_{bonds}) is originated due to the bond stretching inside the molecule. The simplest model to understand bond energy is to treat bonds as springs. The energy of the bond is simply a constant multiplied by the square of the displacement from the equilibrium position, which is a simple spring, as described by Hooke's law.

Molecules can rotate around single bonds, and there is an energy barrier for such rotations. This energy barrier has significant contributions as E_{angles} to E_M. The Vander-Waals energy term is represented as V_{dw} and has the major contribution in E_M. Charge–charge interaction energy becomes relevant when we have polarized centers inside the molecule. Their attractive or repulsive nature constitutes energy (E_{charge}). Different researcher groups around the globe have developed force fields, and they all follow a common scheme for the calculation of molecular mechanics energy, although most have additional terms as well, which generally clubbed and contained under the term $E_{miscelleneous}$.

$$E_M = E_{bonds} + E_{angles} + E_{vdw} + E_{torsion} + E_{charge} + E_{miscelleneous}$$

3.1.2 Molecular Dynamics Algorithm

In MD simulations, the time evolution of a set of interacting particles is followed via the solution of Newton's equations of motion

$$F_i = m_i \frac{d^2 r_i(t)}{dt^2}$$

where $r_i(t) = x_i(t), y_i(t), z_i(t)$ is the position vector of ith particle and F_i is the force acting upon the ith particle at a time and m_i is the mass of the particle.

A 'particle' usually corresponds to an atom; however, it can also represent any distinct entity (e.g., a specific chemical group) that can be conveniently described in terms of a certain interaction law. To integrate the above second-order differential equations, the instantaneous forces acting on the particles and their initial positions and velocities need to be specified. Due to the many-body nature of the problem, the equations of motion are discretized and solved numerically. MD trajectories are defined by both position and velocity vectors and they describe the time evolution of the system in phase space. Accordingly, the positions and velocities are propagated with a finite time interval using numerical integrators, for example, the Verlet algorithm. The position of each particle in space is defined by $r_i(t)$, whereas the velocities $v_i(t)$ determine the kinetic energy and temperature in the system.

As the particles 'move', their trajectories may be displayed and analyzed, providing averaged properties. The dynamic events that may influence the functional properties of the system can be directly traced at the atomic level, making MD especially valuable in molecular biology and materials science. A simple MD algorithm will comprise the following steps:-

Initialize the system
- Read in coordinates at time $t = 0$, set velocities to zero.
- Give final temperature, number of iterations, heating, and cooling rate.
- Criteria for force calculations and setting fixes.

Calculate total potential energy, total energy, and the forces
Write out
Iterate
- Calculate positions for the next timestep.
- Calculate velocities, forces, KE, and temperature.
- Use desired temperature control (thermostat).

(The above step is skipped if constant energy simulation is desired.)
    End iteration (Figure 3.1)

3.1.3 Parameters for MD Simulations

A MD simulation needs a few basic parameters and settings before its implementation-

3.1.3.1 Starting Structure

A MD simulation must begin with a sensible geometry for the structure of interest, and this may be difficult to obtain. Typically, an approximate model is built using available experimental data for crystal structure, and this will then be minimized so that all the bond lengths and bond angles have sensible values. The energy minimization of the structure is also necessary because, if the starting structure is too strained, errors in the MD simulation may accumulate too rapidly, and the structure may blow up.

3.1.3.2 Initial Temperature/Energy

The amount of energy given to the structure depends on the temperature of the system. This energy is divided between kinetic and potential energy parts, just like a ball on a spring will be moving quickly when the spring is relaxed, and stationary

Molecular Dynamics Simulations

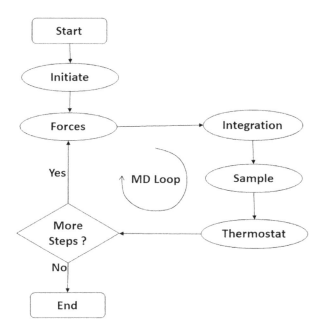

FIGURE 3.1 Functioning of an MD algorithm.

when the spring is fully compressed or fully extended. The energy is divided among all the atoms almost uniformly. The potential function facilitates this distribution. However, if some atoms are getting a considerably higher amount of energy, due to the inappropriate selection of simulation parameters, then such atoms will be 'lost' from the system boundaries.

3.1.3.3 Size of the Timestep

The MD simulation determines the new position of a particle as a function of time from its initial position and velocity. To solve Newton's equations of motion, time is discretized using a timestep. The timestep must be small enough to avoid discretization errors and must be smaller than the inverse of the fastest vibrational frequency. At the same time, it should be large enough for the total simulation duration to access the desired phenomena in a meaningful way. The characteristic time for a MD simulation may be estimated by considering an ordinary infrared spectrum. The peaks in an infrared spectrum correspond to the movements of a molecule. The characteristic time has to be substantially shorter than the fastest of these movements. The shortest period of oscillation in an infrared spectrum is about ten femtoseconds, which indicates a typical value of timestep in many MD simulations.

3.1.3.4 Length of a Simulation

How long should a MD simulation run for? The short answer is 'As long as possible!' Usually, it would be desirable to run it for longer than is feasible, but the choice of a termination time will depend on the information required. A MD simulation begins by giving each atom in a molecule some kinetic energy. This makes the molecule move around, and it is possible to calculate how it moves by solving the Newtonian equations of motion. This is done by analyzing what the molecule is doing and using this to predict what will be happening in a very short time in the future. The calculation is a difficult one, and it takes a lot of computer power to simulate how a molecule will move for a few picoseconds. This is a much shorter time than is generally of interest, but useful information can be gained from short simulations. The length of the simulation is the multiple of the number of timesteps and the number of atoms in the system to be simulated.

3.2 MATERIAL SIMULATION AT THE NANOSCALE AND BULK AND ITS VISUALIZATION

To perform MD simulations, the first requirement is to model a system. There are several ways to do this. We can create a data file containing the complete information of the system separately, or we can prepare it inside our main input file for the MD simulations. However, when we have to prepare a system with some specified geometry of defects or vacancies, then we can simulate it through the tools like Atomsk or Avogadro. These tools have the capability to generate a LAMMPS compatible data file. Therefore, these tools help on the starting step of any MD simulation, which is the creation of the sample material on which we have to perform the 'computational experiment'.

3.2.1 *Atomsk* to Make New Material Structures

Atomsk is a free and open source command-line program dedicated to the creation, manipulation, and conversion of data files for atomic-scale simulations [21]. It can deal with file formats used by many other tools:

- **Visualization and Analysis Tools:** OVITO, Atomeye, VESTA, XCrySDen
- **Ab Initio Tools:** Quantum Espresso, VASP, SIESTA;
- **Molecular Dynamics Simulation Tools:** DL_POLY, LAMMPS;
- **Tunneling Electron Microscope Image Simulation Tools:** Dr. Probe, JEMS.

Atomsk has options for the creation of elementary transformations (duplicating, rotating, deforming, inserting dislocations, merging several systems, creating bicrystals and polycrystals, and so on). These elementary transformation tools can construct and shape a wide variety of atomic systems collectively. In addition, Atomsk can also transform atomic positions, like it can create supercells, cut crystal planes, apply stress, or can insert dislocations.

However, it is important to note that this program is *not* meant to set all simulation parameters, for any particular purpose. Atomsk generally produces a data file containing atom positions only. This data file can be read by the LAMMPS input file for further processing.

Additionally, Atomsk can also perform some elaborate things, like it can produce simultaneous output to multiple formats, can convert a list of files, and can compute the difference between the atomic positions.

3.2.2 LAMMPS MOLECULAR DYNAMICS SIMULATOR

LAMMPS (Large-scale Atomic/Molecular Massively Parallel Simulator) is a classical MD code developed at the Sandia National Laboratory, USA [22]. LAMMPS can model a system of particles in any physical state. It can model atomic, polymeric, biological, solid-state, granular, coarse-grained, or macroscopic systems using a variety of interatomic potentials (force fields) and boundary conditions appropriately. LAMMPS has capabilities to model 2d or 3d systems with only a few particles up to millions or billions. Originally, LAMMPS is designed for parallel computers, but it works efficiently on desktop and laptop computers also.

The current version of LAMMPS is written in C++; however, the initial few versions were written in F77 and F90. LAMMPS is an open-source code and all the versions can be freely downloaded from the LAMMPS website. The unique aspect of LAMMPS is that it can easily be modified or extended with new capabilities, such as new force fields, atom types, boundary conditions, or diagnostics.

LAMMPS works by integrating Newton's equations of motion for a system of atoms or molecules. The interaction models that LAMMPS include are mostly short-range in nature; some long-range models are also included as well.

LAMMPS executes one line at a time, from an input script (text file). Each command makes LAMMPS take some action. Most of the commands in LAMMPS are having some default settings, which can be changed in many ways, if required. An input script of LAMMPS contains mainly following four parts.

3.2.2.1 Initialization

This part of the input script sets the parameters, which are necessary before the creation of atoms. The relevant commands are *units, dimension, newton, processors, boundary, atom_style, atom_modify*.

Also, the kinds of force fields are set by: *pair_style, bond_style, angle_style, dihedral_style, improper_style*.

3.2.2.2 Atom Definition

This part of the input script sets the particle types in the LAMMPS simulation. This can be one of either of the methods below:

- Reading them from a data file or restart file via the *read_data* or *read_restart* commands. Both types of files contain molecular topology details.
- Atoms can be created over a lattice, using commands: *lattice, region, create_box, create_atoms*. Here no initial molecular topology is there.
- The *replicate* command can duplicate the entire set of atoms and make a larger system to be simulated.

3.2.2.3 Settings

After defining atoms and setting the molecular topology, one has to specify other simulation parameters like force field coefficients, output options, and so on.

The selection of appropriate force fields is very important for the appropriate modeling and prediction of any system. It is possible to model a few hundreds to millions of particles in classical MD using phenomenological interatomic and intermolecular potentials. In general, the flexibility, accuracy, transferability, and computational efficiency of the interatomic potentials each have to be carefully considered.

Force field coefficients setting is done with the help of the following commands: *pair_coeff, angle_coeff, bond_coeff, dihedral_coeff, kspace_style, dielectric, special_bonds. improper_coeff.*

Various simulation parameters are set by these commands: *neighbor, neigh_modify, group, timestep, reset_timestep, run_style, min_style, min_modify.*

Fixes impose a variety of boundary conditions, time integration, and diagnostic options. The *fix* command can affect the simulation in a variety of ways by using appropriate keywords.

The computations can be included inside the input script by commands: *compute, compute_modify*, and *variable*.

The output (dump) data lf LAMMPS MD Simulation is controlled by the *thermo, dump,* and *restart* commands.

3.2.2.4 Running a Simulation

Once the input script is completed, the MD simulation can be executed by the *run* command. The minimization of energy before running is very essential. This is done using the *minimize* command. The energy minimization of the system in LAMMPS is performed by iteratively adjusting atom coordinates. These iterations are stopped, when one of the stopping criteria is satisfied. At this point, the system will be in local minimum potential energy. The type of minimization is set by the '*min_style*' command.

3.2.2.5 Ensembles, Constraints, and Boundary Conditions

LAMMPS mostly uses the following three types of statistical ensembles:

1. NPT (keeping the number of atoms, pressure, and temperature constant throughout the simulation)
2. NVE (keeping the number of atoms, volume, and total energy constant throughout the simulation)
3. NVT (keeping the number of atoms, volume, and temperature constant throughout the simulation).

Furthermore, the boundary conditions that are employed in LAMMPS are:

   i. p p p
   ii. p p s
   iii. p f p

where 'p' stands for periodic along the three directions, 'f' is non-periodic and fixed, and 's' stands for non-periodic and shrink wrapped.

The neighbor list is another important aspect that LAMMPS uses, to keep track of adjacent particles. These lists can be optimized for different systems. The spatial-decomposition techniques are used on parallel machines; divide the main simulation system into smaller subsystems, one of which is assigned to each processor. Processors communicate among each other and store the information of "ghost" atom, which border their sub-domain.

3.3 PROPERTY PREDICTION WITH MOLECULAR DYNAMICS

MD offers reliable computational techniques, which have a lot of application in the determination of the physical and thermal properties of materials at the nanoscale. The selection of appropriate force fields also ensures the reliability of the prediction of these simulations.

3.3.1 MECHANICAL PROPERTIES

Properties of nanoscale materials have become an active area of research due to their enhanced physical properties as compared to their respective bulk counterparts. How the mechanical properties are affected by the shape and size of a nanomaterial is very important to understand if these nanomaterials are to be used in applications like MEMS, NEMS, and flexible electronic devices.

The MD method with the use of appropriate force field has already been used for the study of mechanical properties of several nanosystems (e.g. nanoparticles, nanowires, nanotubes, and so on) [23–28]. In the case of metals and alloys, the use of the embedded atom method (EAM) and modified EAM (MEAM) potential has shown considerably good predictability.

Nanoscale materials have a high surface-to-volume ratio, due to which these materials are not having the similar atomic spacing as of the bulk systems. This fact suggests that, before studying any mechanical property computationally, it is necessary to minimize the total energy of the system. LAMMPS allows applying tensile loading in one, two, or three directions. The periodic boundary conditions are to be used in the strained directions of the nanosystem.

LAMMPS along with the visualization tools like OVITO and VMD allows the user to investigate the deformation and yield behavior of the gold nanowires. Figure 3.2 shows the stress–strain graphs [29] for gold nanowires of 4 nm diameter. The simulation parameters used for the simulation are listed in Table 3.2. MD simulations can be done for materials like metals, alloys, metallic glasses, polymers, and solid mixes, of different dimensions, to calculate various mechanical parameters.

The graph initially shows an elastic region where the structural changes that occurred in the shape of the nanowire due to stretching are reversible. The deformation produced in this region is reversible as the elongation in the nanowire takes place only due to the stretching of the bonds between atoms.

The yield point is the place where the elastic region finishes and the plastic region starts. Structural changes in the plastic region are irreversible due to breaking of the

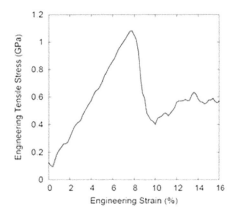

FIGURE 3.2 Stress–strain graph for a gold nanowire of 4 nm diameter [29].

TABLE 3.2
Simulation Parameters for Gold Nanowire Deformation

Parameter	Value
Length	24 nm
Diameter	4 nm
Temperature	100 K
Strain rate	$10^9 s^{-1}$
Timestep	0.005 ps

several bonds. The intercept of the yield point on the y-axis is known as the yield stress while the x-axis intercept is referred to as the yield strain. Young's modulus is defined as the slope of the stress–strain before the yield point. Young's modulus is the stress–strain response of nanomaterials, which looks different from the stress–strain response of bulk materials. Nanomaterials exhibit localized yielding events. These localized yields correspond to the nucleation of a defect, which dissipates internal energy causing a near-instantaneous drop in the stress.

Similarly, the stress–strain graphs for the same nanowire at different temperatures are shown in Figure 3.3. The simulation parameters are the same as given in Table 3.2.

3.3.2 Thermal Properties

MD is by far the most efficient and widely used numerical approach for studying thermodynamic properties in atomic systems. The system behavior is concisely described by the evolution of all the atoms with time and desired properties can be easily extracted. Because of its inherent simplicity in mathematical formulation and numerical algorithm, MD has been widely viewed as an indispensable bridge to connect the microscopic atomic structures to macroscopic material characteristics.

Molecular Dynamics Simulations

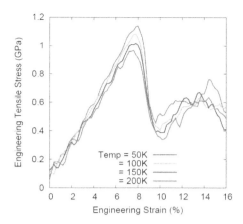

FIGURE 3.3 The dependence of yield stress and strain on temperature. The simulation parameters are the same as in Table 3.2.

In general, it is difficult to obtain an analytical solution that precisely describes the atoms' trajectories. Therefore, we search for a numerical solution for the equations of motion using a time-discretized finite difference method (e.g., Verlet method). The initial atomic positions are defined on its crystal lattice and the starting velocities are decided as per the Boltzmann distribution at the desired temperature. MD uses temperature rescaling by which it can put the average value of temperature very much near the value indicated by the user inside the input file.

The process of a solid changing its state to liquid with the increase in temperature is called melting. In the process of melting, the solid requires energy to transform to liquid and this energy is solely utilized for the change of the phase of the material. Melting is specified as a first-order phase transformation, where the first derivative of Gibbs free energy with temperature and pressure changes discontinuously.

Prediction and monitoring of the melting process in nanoscale devices like nanowires have always been a subject of interest. Many devices like NEMS, MEMS, and flexible electronic devices use nanowires as the communication wires or the network of nanowires serves as an electrode. The thermal stability and its proper monitoring are highly required in such types of devices, which are dependent on the response of nanowires while the working temperature is increasing. These types of studies have always been a challenge for an experimentalist. However, this problem is well suited to be taken up by MD simulations, which had already proven its credibility for the estimation of such properties [30–38].

MD can generate plenty of output parameters, which can be used for the analysis of thermal properties at the nanoscale. These may be the total energy or potential energy of the system, specific heat, mean squared displacement of the particles, Lindemann index, enthalpy, density, volume, etc. One can prepare a LAMMPS Input script for dumping out any of the above parameters from an MD simulation. Figure 3.4 shows the variation of potential energy with increasing temperature for silver nanowires of diameters ranging between 1 and 3 nm. The various parameters used for the simulation are listed in Table 3.3. The sudden decrease of PE indicates

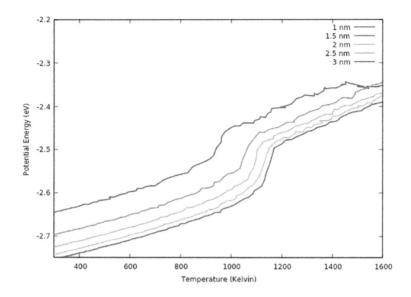

FIGURE 3.4 Variation of potential energy with temperature for the silver nanowires of different dimensions.

TABLE 3.3
Simulation Parameters for the MD Simulation of the Melting of Silver Nanowires

Parameter	Value
Length	20 nm
Diameter	1–3 nm
Temperature range	300–1800 K
Heating rate	3×10^{-3} K/ps
Timestep	0.005 ps

the melting phase transition; the corresponding edge in the curve can be used for the determination of the melting point of that particular nanowire.

The melting point estimated for these silver nanowires is in good agreement with those reported in the literature [39] that justifies the applicability of MD simulation for monitoring of the melting phase transition.

3.4 POST-PROCESSING AND THE VISUALIZATION OF THE OUTPUT DATA

The outputs (dump) of a MD simulation are generally large molecular trajectory or data files, which contain relevant dynamical data to be made usable with the help of graphs and suitable visualization tools. Sometimes, many structure details can only be understood through a proper visualization of the MD dump data. Also, in the validation and analysis process of the output data, we again have to rely substantially on molecular graphics. It is difficult to obtain a spatial understanding of a process or phenomena directly from a trajectory. The use of Computer graphics to understand the output of molecular simulations has now become very much useful and informative. It gives unique information about the microstructure and molecular movements of the simulated system, which never could be obtained after experimentation. However, the success of this visualization output again depends upon the capability of a user in implementing the MD simulations. High-end applications of this category may include interactive virtual reality to offer the scientist the ability to move among the molecular images for a closer look at fast local molecular processes. Fortunately, there are several visualization tools for the display of MD simulation outputs in a way predefined by the user. OVITO and VMD are the most common and easy-to-use visualization tools used in MD.

3.4.1 OPEN VISUALIZATION TOOL (OVITO)

OVITO is the most common data visualization and post-processing tool developed by Dr. Alexander Stukowski [40]. OVITO is an open-source tool, freely available for Windows, Linux, and macOS. OVITO is used for visualizing and analyzing output data from particle-based simulation models, notably MD simulations of solid-state materials. Its focus is on large simulation models based on classical potentials. OVITO has some remarkable features:

- Input and output of various file formats used by MD simulation codes, including LAMMPS.
- Interactive display of large numbers of particles and bonds using hardware-accelerated OpenGL rendering.
- It supports various particle shapes including spheres, cubes, ellipsoids, cylinders, etc.
- Easy-to-use interface with a flexible software design, and a data model that supports an arbitrary number of particle properties.
- High-quality image and movie rendering for publication and a keyframe-based animation system. At any stage of simulation, animation/images can be published with tens of additional parameters.
- A powerful Python-based scripting interface that allows one to automate analysis and visualization tasks and extend OVITO with new functions.
- Nondestructive data manipulation and analysis functions are arranged in a data flow pipeline by the user. A data flow pipeline enables the user to manipulate the data non-destructively. Users can change the parameters at any time and see the impacts in real-time. A data flow pipeline is constituted by various modifications given in OVITO.

OVITO can directly read the LAMMPS trajectory file (.lammpstrj), which can be dumped by a LAMMPS input script.

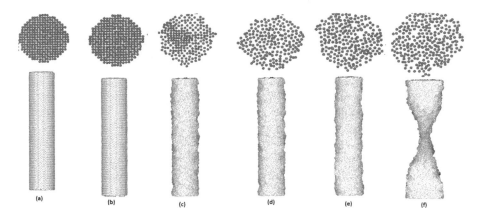

FIGURE 3.5 Effect of heating on a silver nanowire of 2 nm diameter, at (a) 300 K, (b) 500 K, (c) 800 K, (d) 1000 K, (e) 1500 K, and (f) 1800 K, rendered using OVITO.

3.4.2 Visual Molecular Dynamics (VMD)

VMD (Visual Molecular Dynamics) is an open-source tool for the three Dimensional modeling, visualization, and analysis of molecular systems [41]. It is a powerful visualization tool and a molecular modeling and visualization computer program. VMD is developed as mainly a tool to view and analyze the results of MD simulations, for working with volumetric data, sequence data, and arbitrary graphics objects. Molecular scenes can be exported in a variety of formats. Users can run their own Tcl and Python scripts within VMD as it includes embedded Tcl and Python interpreters.

The *topotool* utility of VMD can be used to create the data file for any molecular system which can be further used in any LAMMPS input script. Firstly we create the LAMMPS trajectory files with the help of the LAMMPS input script. Thereafter, the LAMMPS trajectory file can be used for the determination of the structure (Figure 3.6).

VMD accepts several types of input file formats like PDB (Protein Data Bank) files having .pdb file extension. LAMMPS can generate .*lammpstrj* files, which can directly be visualized by VMD, to render a movie or high-resolution image. Figure 3.6 displays the various stages of deformation in a gold nanowire, when the tensile loading is applicable along its length. The simulation parameters are as given in Table 3.1.

3.5 MACHINE LEARNING FOR PERFORMANCE ENHANCEMENT OF MOLECULAR DYNAMICS SIMULATIONS

In the fields of materials science, physics, chemistry, and bioengineering, the parallel computing techniques provide the needed performance enhancement for carrying out lengthy simulations of complex and big systems. Suppose, in a typical MD simulation of a nanosystem, $\approx$1 nanosecond of dynamics of $\approx$500 particles on a single processor takes approximately 12 hours of runtime, which is comparatively large enough. On the other hand, using OpenMP shared memory and multiple cores, the runtime may be reduced up to approximately 1 hour or even less than that. We expect

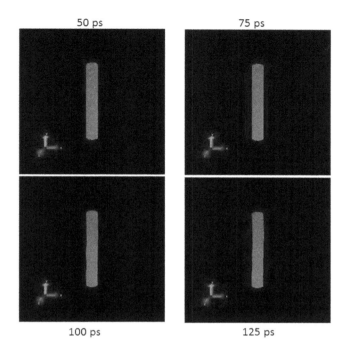

FIGURE 3.6 The VMD snapshots showing the deformation of the gold nanowire with time [29].

a further reduction in the runtime by using a hybrid OpenMP/MPI approach. All these techniques enable state-of-the-art research with simulations that can explore the dynamics of constituent particles with fewer approximations, more accurate potentials, and hundreds of nanoseconds over a wider range of physical parameters. Even after the employment of the optimal parallelization techniques which suit the size and complexity of the system, simulations of complex scientific phenomena remain time consuming. However, several researchers have already started implementing the Machine Learning (ML) approaches to enhance the performance and predictability of MD simulations [42–50].

In classical MD simulations, the physical interactions between atoms are estimated with the help of an appropriate force field, which contains a large number of parameters for each constituent particle. The time evolution of these particles is recorded with the help of coordinates and energy trajectories. In addition, the ensemble averages are determined to compare various properties with experimentally known values for the validation of MD simulations and the implemented force fields. These properties can be density, free energy of solvation, heat capacity, self-diffusion coefficient, viscosity, radial distribution function, radius of gyration, etc. However, averaging the trajectories may discard the potentially valuable information related to kinetics and the relative population of conformational states. Recently, kinetic modeling approaches like the Markov state model (MSM) have emerged as a tool to harness the information of time from the stored MD trajectories [51].

However, the information from a separate set of MD simulations can be clubbed with this approach. The construction of an MSM is a two-step process. In the discretization step, the snapshots of the trajectories are structurally clustered into microstates and in the metastable step, microstates are kinetically clustered.

With the converged kinetic models, the total time of the simulation is in the range of microseconds or milliseconds, typically. In the recent past, various researchers have proposed different ML approaches for kinetic modeling [52–55].

ML methods can be applied in several ways for improving the accuracy and predictability of empirical force fields. ML models can be trained to learn the potential-energy profiles such that they may be able to replace the force field entirely. However, the big challenge for the method is the generation of a diversified training set in terms of chemical parameters and conformational space.

3.5.1 Methodology of Applying Machine Learning in Molecular Dynamics Simulations

ML has successfully been used for the enhancement of the overall performance and usability of a system at the nanoscale. Figure 3.5 shows the overview of the typical implemented methodology. First, the attributes of the media conditions that define the system are fed to the simulation framework. With the help of these inputs, the MD simulation is launched on the high-performance computing (HPC) cluster. Simultaneously, the same inputs are also given to the prediction module for ML. The output of the MD simulation characterizes the structure of the system near interfaces. Error handler aborts the program and displays appropriate error messages when a simulation fails due to any pre-defined criteria. In addition, at the end of the simulation run, parameter values are saved for future retraining of the ML model (Figure 3.7).

3.5.2 Machine Learning and Related Approaches to Enhance Sampling

ML not only learns from the provided data, but it has the capability of data sampling. Especially, when the system size is large and complex, a heavy amount of dump data is used for analysis, making the data processing slow. In such situations, unbiased

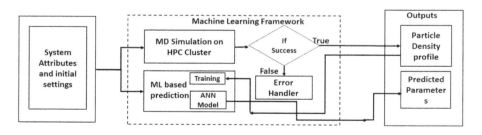

FIGURE 3.7 Schematic representation of the implementation of machine learning in molecular dynamics.

Molecular Dynamics Simulations

sampling, even by the use of the best computing resources, does not seem possible. On the other hand, the success of machine learning techniques largely depends upon the abundance of the data. However, we may not have the relevant data to train the ML model; one could train the model by enhanced data sampling methods. Again, these methods themselves require an estimate of reaction coordinates (RC). However, a wrong choice of RC can mislead the model which is being trained. To avoid any such instances, iteration is done between the rounds of sampling. With every round of sampling, we expect improved reaction co-ordinates, which may be used to perform better and enhanced sampling and so on.

One can use ML to not just learn from given data but actually generate statistically accurate information when the underlying processes are so slow that they simply cannot be sampled in unbiased MD even with the best available computing resources. Essentially, the success of ML highly relies on the abundance of data. However, if the events of interest are rare, one faces a paucity of relevant data to train the ML upon. One could train ML models on data from enhanced sampling methods, but these methods themselves need an estimate of the RCs, and a wrong choice of RCs used to generate the data could mislead the ML being trained upon it. One solution to this dilemma is to iterate between rounds of sampling and ML, where every ML round generates a progressively improved RC, which is used to perform better-enhanced sampling, and so on.

The molecular enhanced sampling with autoencoders (MESA) approach uses an Artificial Neural Network (ANN) with a non-linear encoder and decoder to learn the RC from input data, which itself is generated through umbrella sampling along trial RCs [56]. Every round of ML leads to an improved RC along which new umbrella sampling is performed. The iterations are continued until the free energy from umbrella sampling no longer varies with further iterations.

Boltzmann Generators are a very recent deep learning-based approach that learns the equilibrium probability P without resorting to running long trajectories [57]. It leverages recent advances in probabilistic generative modeling in which invertible coordinate transformations mapping x onto a random variable x_0 whose distribution is straightforward to sample are learned [58]. Using x_0 together with the fact that x follows the Boltzmann distribution, the ANN can be trained (in principle) without using maximum likelihood on a pre-existing dataset.

3.6 CONCLUSION

The MD simulation technique is a powerful tool to see the atomic insights of a sample. MD simulations can be used to understand the properties of system particles, for the modeling of their structure, and to understand interactions between them at the microscopic level.

Combined with Machine Learning approaches, it enhances its performance and predictability. There are several approaches in machine learning which can be used with MD. A large amount of data can be used appropriately to train the Machine Learning module appropriately, and later can be used for the prediction of various related parameters. The ML approach can also be used effectively for the sampling of the large volume MD simulation data.

REFERENCES

1. Alder, B. J., & Wainwright, T. E. (1959). Studies in molecular dynamics. I. General method. *The Journal of Chemical Physics*, *31*(2), 459–466. https://doi.org/10.1063/1.1730376.
2. Rahman, A. (1964). Correlations in the motion of atoms in liquid argon. *Physical Review*, *136*(2A), A405–A411. https://doi.org/10.1103/PhysRev.136.A405.
3. Dong, Y., Li, Q., & Martini, A. (2013). Molecular dynamics simulation of atomic friction: A review and guide. *Journal of Vacuum Science & Technology A*, *31*(3), 030801. https://doi.org/10.1116/1.4794357.
4. Shimizu, J., Eda, H., Yoritsune, M., & Ohmura, E. (1998). Molecular dynamics simulation of friction on the atomic scale. *Nanotechnology*, *9*(2), 118–123. https://doi.org/10.1088/0957-4484/9/2/014.
5. Sun, W., & Mai, Y.-W. (2019). Molecular dynamics simulations of friction forces between silica nanospheres. *Computational Materials Science*, *162*, 96–110. https://doi.org/10.1016/j.commatsci.2019.02.025.
6. Eder, S., Vorlaufer, G., Ilincic, S., & Betz, G. (2007, October 1). Simulation of wear processes using molecular dynamics (MD) simulations. https://doi.org/10.13140/2.1.4149.5680.
7. James, S., & Sundaram, M. M. (2015). Molecular dynamics simulation study of tool wear in vibration assisted nano-impact-machining by loose abrasives. *Journal of Micro and Nano-Manufacturing*, *3*(1). https://doi.org/10.1115/1.4028782.
8. Zhou, Y. L., Tao, X. M., Hou, Q., & Ouyang, Y. F. (2018). Molecular dynamics simulations of atoms diffusion in solid. *Diffusion Foundations*; Trans Tech Publications Ltd. https://doi.org/10.4028/www.scientific.net/DF.15.51.
9. Sun, W., & Wang, H. (2020). Molecular dynamics simulation of diffusion coefficients between different types of rejuvenator and aged asphalt binder. *International Journal of Pavement Engineering*, *21*(8), 966–976. https://doi.org/10.1080/10298436.2019.1650927.
10. (n.d.) Molecular dynamics simulations of adsorption and diffusion of gases in silicon-carbide nanotubes. *The Journal of Chemical Physics*, *132*(1). Retrieved February 7, 2021, from https://aip.scitation.org/doi/abs/10.1063/1.3284542?journalCode=jcp.
11. Guisbiers, G. (2019). Advances in thermodynamic modelling of nanoparticles. *Advances in Physics: X*, *4*(1), 1668299. https://doi.org/10.1080/23746149.2019.1668299.
12. Lv, Y. J., & Chen, M. (2011). Thermophysical properties of undercooled alloys: An overview of the molecular simulation approaches. *International Journal of Molecular Sciences*, *12*(1), 278–316. https://doi.org/10.3390/ijms12010278.
13. al-Dosari, M. S., & Walker, D. G. (2011). Thermal properties of yttrium aluminum garnett from molecular dynamics simulations. *ASME/JSME 2011 8th Thermal Engineering Joint Conference*, T10046. https://doi.org/10.1115/AJTEC2011-44343.
14. Gafner, S. L., Redel, L. V., & Gafner, Yu. Ya. (2012). Molecular-dynamics simulation of the heat capacity for nickel and copper clusters: Shape and size effects. *Journal of Experimental and Theoretical Physics*, *114*(3), 428–439. https://doi.org/10.1134/S1063776112010116.
15. Ryu, S., & Cai, W. (2008). Comparison of thermal properties predicted by interatomic potential models. *Modelling and Simulation in Materials Science and Engineering*, *16*(8), 085005. https://doi.org/10.1088/0965-0393/16/8/085005.
16. (2017) Molecular dynamics simulation of the adsorption and aggregation of ionic surfactants at liquid–solid interfaces. *The Journal of Physical Chemistry C*, *121*(46), 25908–25920. https://pubs.acs.org/doi/abs/10.1021/acs..jpcc.7b09466.
17. Krasnikov, V. S., & Mayer, A. E. (2016). Melting of aluminum with ideal or defect lattice: Molecular dynamics simulations with accounting of electronic heat conductivity. *Journal of Physics: Conference Series*, *774*, 012016. https://doi.org/10.1088/1742-6596/774/1/012016.

18. Liu, W., Chen, P., Qiu, R., Khan, M., Liu, J., Hou, M., & Duan, J. (2017). A molecular dynamics simulation study of irradiation induced defects in gold nanowire. *Nuclear Instruments and Methods in Physics Research Section B: Beam Interactions with Materials and Atoms*, *405*, 22–30. https://doi.org/10.1016/j.nimb.2017.05.016.
19. Adcock, S. A., & McCammon, J. A. (2006). Molecular dynamics: Survey of methods for simulating the activity of proteins. *Chemical Reviews*, *106*(5), 1589–1615. https://doi.org/10.1021/cr040426m.
20. Yoo, J., Winogradoff, D., & Aksimentiev, A. (2020). Molecular dynamics simulations of DNA–DNA and DNA–protein interactions. *Current Opinion in Structural Biology*, *64*, 88–96. https://doi.org/10.1016/j.sbi.2020.06.007.
21. Hirel, P. (2015). Atomsk: A tool for manipulating and converting atomic data files. *Computer Physics Communications*, *197*, 212–219. https://doi.org/10.1016/j.cpc.2015.07.012.
22. Plimpton, S. (1995). Fast parallel algorithms for short-range molecular dynamics. *Journal of Computational Physics*, *117*(1), 1–19. https://doi.org/10.1006/jcph.1995.1039.
23. Liu, W. K., Karpov, E. G., Zhang, S., & Park, H. S. (May 2004) An introduction to computational nanomechanics and materials. *Computer Methods in Applied Mechanics and Engineering*, *193*(17–20), 1529–1578. https://doi.org/10.1016/j.cma.2003.12.008.
24. Gall, K., Diao, J., & Dunn, M. L. (December 2004) The strength of gold nanowires. *Nano Letters*, *4*(12), 2431–2436. https://doi.org/10.1021/nl048456s.
25. Cheng, H.-C., Yu, C.-F., & Chen, W.-H. (2014). Size, temperature, and strain-rate dependence on tensile mechanical behaviors of Ni_3Sn_4 intermetallic compound using molecular dynamics simulation. *Journal of Nanomaterials*, *2014*, 1–17. https://doi.org/10.1155/2014/214510.
26. Reischl, B., Rohl, A. L., Kuronen, A., & Nordlund, K. (2017). Atomistic simulation of the measurement of mechanical properties of gold nanorods by AFM. *Scientific Reports*, *7*(1), 16257. https://doi.org/10.1038/s41598-017-16460-9.
27. Xiao, L., Zhang, J., Zhu, Y., Shi, T., & Liao, G. (2019). Molecular dynamics simulation of the tensile mechanical behaviors of axial torsional copper nanorod. *Journal of Nanoparticle Research*, *21*(8), 169. https://doi.org/10.1007/s11051-019-4609-z.
28. Salmankhani, A., Karami, Z., Hamed Mashhadzadeh, A., Saeb, M. R., Fierro, V., & Celzard, A. (2020). Mechanical properties of C3N nanotubes from molecular dynamics simulation studies. *Nanomaterials*, *10*(5), 894. https://doi.org/10.3390/nano10050894.
29. Joshi, S. K., Pandey, K., Singh, S. K., & Dubey, S. (2019). Molecular dynamics simulations of deformation behaviour of gold nanowires. *Journal of Nanotechnology*. https://doi.org/10.1155/2019/5710749.
30. Sar, D. K., & Nanda, K. K. (2010). Melting and superheating of nanowires—A nanotube approach. *Nanotechnology*, *21*(20), 205701. https://doi.org/10.1088/0957-4484/21/20/205701.
31. Chen, H.-L., Ju, S.-P., Wang, S.-L., Pan, C.-T., & Huang, C.-W. (2016). Size-dependent thermal behaviors of 5-fold twinned silver nanowires: A computational study. *The Journal of Physical Chemistry C*, *120*(23), 12840–12849. https://doi.org/10.1021/acs.jpcc.6b02412.
32. Hui, L., Wang, B. L., Wang, J. L., & Wang, G. H. (2004). Melting behavior of one-dimensional zirconium nanowire. *The Journal of Chemical Physics*, *120*(7), 3431–3438. https://doi.org/10.1063/1.1640613.
33. Alarifi, H. A., Atiş, M., Özdoğan, C., Hu, A., Yavuz, M., & Zhou, Y. (2013). Determination of complete melting and surface premelting points of silver nanoparticles by molecular dynamics simulation. *The Journal of Physical Chemistry C*, *117*(23), 12289–12298. https://doi.org/10.1021/jp311541c.

34. Cheng, D., & Cao, D. (2008). Structural transition and melting of onion-ring Pd–Pt bimetallic clusters. *Chemical Physics Letters*, *461*(1–3), 71–76. https://doi.org/10.1016/j.cplett.2008.06.062
35. Ryu, S., & Cai, W. (2008). Comparison of thermal properties predicted by interatomic potential models. *Modelling and Simulation in Materials Science and Engineering*, *16*(8), 085005. https://doi.org/10.1088/0965-0393/16/8/085005.
36. Gang, C., Peng, Z., & HongWei, L. (2013). Analysis of Pd-Ni nanobelts melting process using molecular dynamics simulation. *Journal of Nanomaterials*, *2013*, 1–7. https://doi.org/10.1155/2013/486527.
37. Essajai, R., & Hassanain, N. (2018). Molecular dynamics study of melting properties of gold nanorods. *Journal of Molecular Liquids*, *261*, 402–410. https://doi.org/10.1016/j.molliq.2018.04.051.
38. Kowaki, Y., Harada, A., Shimojo, F., & Hoshino, K. (2007). Radius dependence of the melting temperature of single-walled carbon nanotubes: Molecular-dynamics simulations. *Journal of Physics: Condensed Matter*, *19*(43), 436224. https://doi.org/10.1088/0953-8984/19/43/436224.
39. Gordon, E., Karabulin, A., Matyushenko, V., Sizov, V., & Khodos, I. (2014). Stability and structure of nanowires grown from silver, copper and their alloys by laser ablation into superfluid helium. *Physical Chemistry Chemical Physics*, *16*(46), 25229–25233. https://doi.org/10.1039/c4cp03471f.
40. Stukowski, A. (2009). Visualization and analysis of atomistic simulation data with OVITO–the Open Visualization Tool. *Modelling and Simulation in Materials Science and Engineering*, *18*(1), 015012. https://doi.org/10.1088/0965-0393/18/1/015012.
41. Humphrey, W., Dalke, A., & Schulten, K. (1996). VMD: Visual molecular dynamics. *Journal of Molecular Graphics*, *14*(1), 33–38. https://doi.org/10.1016/0263-7855(96)00018-5.
42. Gastegger, M., Behler, J., & Marquetand, P. (2017). Machine learning molecular dynamics for the simulation of infrared spectra. *Chemical Science*, *8*(10), 6924–6935. https://doi.org/10.1039/C7SC02267K
43. Gebhardt, J., Kiesel, M., Riniker, S., & Hansen, N. (2020). Combining molecular dynamics and machine learning to predict self-solvation free energies and limiting activity coefficients. *Journal of Chemical Information and Modeling*, *60*(11), 5319–5330. https://doi.org/10.1021/acs.jcim.0c00479.
44. Botu, V., & Ramprasad, R. (2015) Adaptive machine learning framework to accelerate ab initio molecular dynamics. *International Journal of Quantum Chemistry*, *115*(16), 1074–1083.
45. Noé, F. (2020). Machine learning for molecular dynamics on long timescales. In K. T. Schütt, S. Chmiela, O. A. von Lilienfeld, A. Tkatchenko, K. Tsuda, & K.-R. Müller (Eds.), *Machine Learning Meets Quantum Physics* (pp. 331–372). Springer International Publishing. https://doi.org/10.1007/978-3-030-40245-7_16.
46. Noé, F., Tkatchenko, A., Müller, K.-R., & Clementi, C. (2020). Machine learning for molecular simulation. *Annual Review of Physical Chemistry*, *71*(1), 361–390. https://doi.org/10.1146/annurev-physchem-042018-052331.
47. Bennett, W. D., He, S., Bilodeau, C. L., Jones, D., Sun, D., Kim, H., Allen, J., Lightstone, F. C., & Ingólfsson, H. I. (2020). Predicting small molecule transfer free energies by combining molecular dynamics simulations and deep learning. *Journal of Chemical Information and Modeling*, *60*(11), 5375–5381.
48. Goh, G. B., Hodas, N. O., & Vishnu, A. (2017). Deep learning for computational chemistry. *Journal of Computational Chemistry*, *38*(16), 1291–1307.
49. Vijayaraghavan, V., Garg, A., Wong, C. H., & Tai, K. (2014). Estimation of mechanical properties of nanomaterials using artificial intelligence methods. *Applied Physics A*, *116*(3), 1099–1107.

50. Wang, Y., Lamim Ribeiro, J. M., & Tiwary, P. (2020). Machine learning approaches for analyzing and enhancing molecular dynamics simulations. *Current Opinion in Structural Biology*, *61*, 139–145. https://doi.org/10.1016/j.sbi.2019.12.016.
51. Huang, X., & De Fabritiis, G. (2014). Understanding molecular recognition by kinetic network models constructed from molecular dynamics simulations. In G. R. Bowman, V. S. Pande, & F. Noé (Eds.), *An Introduction to Markov State Models and Their Application to Long Timescale Molecular Simulation* (pp. 107–114). Springer: Netherlands. https://doi.org/10.1007/978-94-007-7606-7_9.
52. Wehmeyer, C., & Noé, F. (2018). Time-lagged autoencoders: Deep learning of slow collective variables for molecular kinetics. *The Journal of Chemical Physics*, *148*(24), 241703. https://doi.org/10.1063/1.5011399.
53. Chen, W., Sidky, H., & Ferguson, A. L. (2019). Nonlinear discovery of slow molecular modes using state-free reversible VAMPnets. *The Journal of Chemical Physics*, *150*(21), 214114. https://doi.org/10.1063/1.5092521.
54. Mardt, A., Pasquali, L., Wu, H., & Noé, F. (2018). VAMPnets for deep learning of molecular kinetics. *Nature Communications*, *9*(1), 5. https://doi.org/10.1038/s41467-017-02388-1.
55. Hernández, C. X., Wayment-Steele, H. K., Sultan, M. M., Husic, B. E., & Pande, V. S. (2018). Variational encoding of complex dynamics. *Physical Review E*, *97*(6), 062412. https://doi.org/10.1103/PhysRevE.97.062412.
56. Chen, W., & Ferguson, A. L. (2018). Molecular enhanced sampling with autoencoders: on-the-fly collective variable discovery and accelerated free energy landscape exploration. *The Journal of Computational Chemistry*, *39*, 2079–2102.
57. Noé, F., Olsson, S., Köhler, J., & Wu, H. (2019). Boltzmann generators: Sampling equilibrium states of many-body systems with deep learning. *Science*, *365*(6457), eaaw1147. https://doi.org/10.1126/science.aaw1147.
58. Dinh, L., Sohl-Dickstein, J., & Bengio, S. (2017). Density estimation using Real NVP. *ArXiv:*1605.08803 *[Cs, Stat]*. http://arxiv.org/abs/1605.08803.

4 Desirability Approach-Based Optimization of Process Parameters in Turning of Aluminum Matrix Composites

P.V. Rajesh
Saranathan College of Engineering

A. Saravanan
National Institute of Technology

CONTENTS

4.1 Introduction: Background and Driving Forces 72
4.2 Selection of Materials 74
    4.2.1 Matrix 74
    4.2.2 Reinforcement 75
4.3 Experimental Procedure 75
    4.3.1 Fabrication Method 75
    4.3.2 Machining Process 76
4.4 Conduction of Tests and Obtained Results 79
    4.4.1 Physical Tests 79
        4.4.1.1 Surface Roughness Test 79
        4.4.1.2 Roundness Test 80
    4.4.2 Mechanical Tests 83
        4.4.2.1 Hardness Test 83
        4.4.2.2 Impact Test 83
4.5 Machine Learning Inspired Optimization 86
4.6 Conclusion 95
References 98

DOI: 10.1201/9781003121954-4

4.1 INTRODUCTION: BACKGROUND AND DRIVING FORCES

Composite is indeed a fascinating word, heard more often nowadays. The term 'composite' means joining two or more materials together at a desirable proportion. At this point, it differs completely from an alloy, in which two or more metals should be mixed at a specific proportion. The composite will not give birth to a new material like alloys, in which the mixing of two or more metals produces an entirely different metal. In composites, the joining materials may or may not be metals and there is no hard and fast rule in deciding their proportion. In other words, if an alloy is created, the physical and chemical properties are altered, when compared to base metals, whereas if a composite is created, the mechanical and technological properties are going to be different from the base materials.

The composites' constituents can be either metals or non-metals such as ceramics, polymer resins, and plastics. The choice of major and minor constituents in composites are decided based on applications or desirable property requirements, which lead to favorable outcomes [1]. Normally, a composite consists of two parts, viz., matrix and reinforcements. Matrix, the major constituent with a larger proportion may be a metal, polymer, or ceramic. Similarly, reinforcement, the minor constituent with a comparatively lesser presence, may be in the form of powders, short whiskers, or lengthy fibers.

A metal matrix composite, in which a base metal alloy is reinforced with suitable materials, is finding an increased usage in the fields like automobile structures, ship hulls, aircraft cabins, etc., to name a few. Praveen Kittali and his associates have reviewed the influences of different reinforcements on the mechanical as well as wear behavior of aluminum metal matrix composites (AMMCs) [2]. MMCs are increasingly preferred for their good alloying ability, superior strength to weight ratio, and perfect susceptibility to heat treatment. They also discussed various manufacturing techniques like pressure infiltration, powder metallurgy, stir casting, and mechanical alloying for fabricating MMCs. Most importantly, aluminum matrix composites (AMCs) are both researcher-friendly and user-friendly as far as their behavioral and functional aspects are concerned. Amol Mali et al. elaborated the usage of aluminum alloys in the automobile, aviation, marine, biotechnology, and mining industries, owing to their low density, attractive mechanical properties, and low coefficient of thermal expansion [3].

The majority of the studies carried out worldwide are concentrating vehemently on the evaluation and experimental investigation of the physical, mechanical, and tribological properties of AMCs reinforced with ceramic particulates or powders.

Due to the excellent formability of aluminum alloy and its composites, which is reflected in their malleability, ductility, and softness, several operations like metal removal, metal joining, and metal forming are carried out. Michel Oluwatosin Bodunrin and K.K. Alaneme conducted a detailed review and presented the same on the comprehensive reinforcement methodologies, corrosion, tribological and mechanical behaviors of aluminum hybrid metal matrix composites. They explained different fabrication techniques, various reinforcement materials, physical and chemical phenomena to be considered, and a wide range of mechanical, technological, and tribological attributes to be investigated in their paper [4].

Vengadesh D. and his colleagues have elaborated on recent composite technologies and surface behavior of AMMCs with various reinforcement strategies in their review paper. Especially, among the materials of tribological importance, AMMCs have received extensive attention due to fundamental and applied reasons [5]. Weight reduction and cost reduction are the important areas to be addressed while selecting AMCs.

Artificial Intelligence (AI) is the gift of the millennium. AI is the ability exhibited by a computational device, most preferably a computer, to think and act by itself without human assistance, control, or intervention. In other words, the computer need not wait for an operator to give instructions regarding what to do, when to do, and how to do. Yet, a computer or computerized device cannot perform without any precedence. So, AI means that the computer should work on pre-coded algorithms and past instructions. It should retrieve past recorded data, store some time before, and perform accomplished tasks smartly, according to the current situation and scenario. A fine example of an AI-enabled device is a Robot. It is reprogrammable and multifunctional and can carry out instructions with ease. The mid-2010s saw a surge in the dependency of such AI implemented systems for reducing manual labor to a greater extent. Nowadays, computer assistance with AI capabilities is needed in a variety of fields across a broader spectrum.

Most importantly process and product industries belonging to manufacturing, production, and mechanical domains are utilizing AI extensively. Let us take the curious case of manufacturing. In this field, usage of AI-based equipment and devices is almost everywhere starting from sensor-enabled metrology and measurements, new product design and development, modification of an existing product to suit customer needs, modeling, engineering, production control, analysis, and inspection, assembly to transportation, thus paving the way to Smart Manufacturing [6].

The most common term, which is uttered frequently in the manufacturing sector, is Internet-of-Things (IoT), nowadays. It is the most important pillar of AI. As IoT is more diversified in recent years, the production fraternity coined a more specific abbreviation IIoT meaning Industrial Internet of Things. IIoT is profoundly used for automatic monitoring and control of manufacturing processes, automatic testing and evaluation of manufacturing components, automatic analysis and prediction of values of material properties, and so on. Hence, it is suitably called Industry 4.0.

Further classification of AI and IIoT to be more generically applied across different streams and disciplines gave rise to Data Science, a boon of the 21st century. Data science is a multi-disciplinary philosophy that uses scientific methods, mathematical calculations, the relationship between different processes, algorithms, equations, and systems to extract solutions and precise predictions from many structural and unorganized data. Its broader appeal led to creating different subsets like Deep Learning, Data Mining, Cloud Computing, Big Data Analytics that are increasingly used in manufacturing functions and operations in modern times [7,8].

Machine Learning is a wider sub-sect of Data Science that builds an analytical model automatically. It is based on the idea that computer systems or automated motorized components can learn from previous data, identify or create similar patterns, make decisions and solve complex problems with less or negligible human effort [9].

Optimization in manufacturing and testing is the most recent computational technique to select the most viable composite specimen from a set of alternatives. A series of different specimens are made by varying process parameters involved in a particular process. Particularly, Machine Learning (ML) and Data Science enabled numerical optimization techniques, which employ Artificial Intelligence (AI) allow software programs and applications to become more precise in predicting various mechanical and tribological properties without being taught to do so [10]. Machine Learning algorithms utilize standard historical data from past experiments as input factors to predict new output values, even without conducting tests [11].

Response Surface Methodology (RSM) is one such technique that uses the Linear Regression Model (LRM) to establish relevance between independent process parameters and test results, depending upon them [12].

4.2 SELECTION OF MATERIALS

A Composite is made up of essentially two parts: Matrix and Reinforcement. Matrix is the outer part that provides the necessary bond as an outer layer, whereas reinforcement is embedded or thrust inside the matrix to provide strength and hardness to the composite [13].

4.2.1 Matrix

Al 6061 is a precipitation-hardened wrought aluminum alloy chosen to be the metallic matrix due to its formidable surface finish and good corrosion resistance. With nearly 97% aluminum and magnesium and silicon accounting for 0.97% and 0.62% respectively as major contributors, Al 6061 is sometimes called Al-Mg-Si alloy. Bharath V. et al. explained the preparation of Al 6061-Al_2O_3 MMCs by Stir Casting. This paper postulated the advantages of using Al 6061 as a matrix material, in terms of its flexibility, physical and chemical properties, formability, and excellent corrosion resistance [14].

The chemical composition of Al 6061, obtained from elemental analysis done by Energy Dispersive Atomic Spectroscopy (EDAS) is given in Table 4.1.

TABLE 4.1
Proportion of Chemical Constituents in Al 6061

Constituents	Percentage (%)
Aluminum (Al)	97.66
Magnesium (Mg)	0.97
Silicon (Si)	0.62
Copper (Cu)	0.39
Iron (Fe)	0.13
Manganese (Mn)	0.11
Chromium (Cr)	0.08
Others (Total)	0.04

4.2.2 Reinforcement

The ceramic material, selected as reinforcement inside the matrix is aluminum oxide alias alumina (Al_2O_3). Valibeygloo N. discussed the microstructural and mechanical properties of Al-4.5 wt.% Cu reinforced with alumina particles. This paper gives detailed information about the properties, abrasiveness, refractoriness, and possible applications of the widely used ceramic, alumina [15].

Being widely regarded as one of the finest ceramics, alumina provides fair refractoriness, hardness, wear-resistance, and better miscibility with aluminum, when compared to other ceramics [16]. Table 4.2 indicates some common properties of alumina.

4.3 EXPERIMENTAL PROCEDURE

In this section, the fabrication method for producing composite specimens, the machining process involved for shaping those specimens, and various tests conducted to measure surface and bulk properties are discussed in detail.

4.3.1 Fabrication Method

Stir casting is one of the most commonly used techniques for producing aluminum-based MMCs. In this method, Al 6061 rods are placed in a crucible, made of graphite, and heated inside a coal-fired furnace at elevated temperatures for about half an hour, till they are melted and attain a semi-solid state [17]. Then, preheated alumina powder is poured carefully at a constant pour rate into the molten metal and stirred thoroughly and continuously with either a manual stirrer or motorized stirrer for nearly 2 minutes to ensure the uniform distribution of reinforcement particles inside the aluminum matrix. Finally, the composite is poured into a mold cavity, that resembles a cylindrical shape with 150 mm length and 30 mm diameter.

TABLE 4.2
General Properties of Al_2O_3

Properties	Values	Units
Density	3.98	g/cm^3
Melting point	2300	Oc
Vickers hardness	1560	–
Fracture toughness	4.9	MPa $\sqrt{m}$
Elastic modulus	300	GPa
Tensile strength	210	MPa
Thermal conductivity	21	W/mK
Coefficient of thermal expansion	9	m/°C

The specimens are prepared as per volume-based ratio by varying the percentage weight of reinforcement materials [18]. The different compositions of composite fabricated using the Stir Casting process are:

- With 5% of alumina
- With 10% of alumina
- With 15% of alumina

Waleed T. Rashid performed a study on how the particle size and stirring speed affect the hardness of AMCs. Here, aluminum (Al6063) was reinforced with Alumina. The results indicated that increasing the stirring speed improves the hardness (because of the uniform distribution), whereas a substantial Increase in the size, increases the hardness [19].

The process is shown in Figure 4.1.

4.3.2 Machining Process

The machining process preferred here is Turning. Turning is a metal removal process done to reduce the diameter, which reduces the circumference of the composite specimens with a circular cross-section. Initially, the composite rod to be turned is fixed to the headstock spindle of the center lathe and made to rotate at a reasonable spindle speed. The tool, generally High-Speed Steel (HSS) is fixed to the tool post with adjusted height. The carriage is positioned at a constant feed rate. Then, the tool is reciprocally moved to contact the rotating workpiece. Finally, the diameter is reduced by applying a suitable depth of cut [20]. The setup for the turning process is shown in Figure 4.2.

The turned composite specimens are indicated in Figure 4.3.

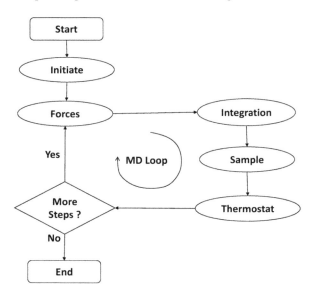

FIGURE 4.1 Stir casting.

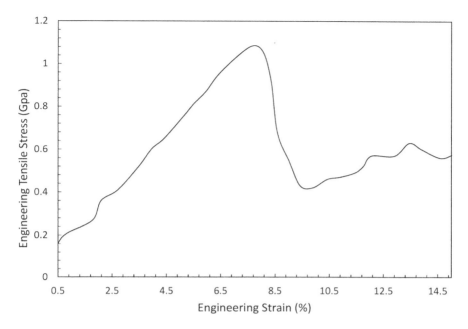

FIGURE 4.2 Turning process setup.

The process parameters used in this process are spindle speed, depth of cut, and feed rate [21]. While spindle speed and depth of cut are variable parameters, the feed rate is constant. Puneet Bansal et al., in their paper, discussed the effect of primary turning parameters such as spindle speed, feed, and depth of cut on material

FIGURE 4.3 Turned specimens.

TABLE 4.3
Factors and Their Levels in the Turning Process

Symbol	Factors	Experimental Levels	
		Low Level	High Level
A	Composition of Al_2O_3 (wt. %)	5	15
B	Spindle speed (rpm)	220	460
C	Depth of cut (mm)	0.2	0.6

properties like tool wear, surface roughness, tensile strength, hardness, and metal removal rate of alumina-reinforced aluminum composites. Their detailed investigation revealed that the above-mentioned properties had a good enhancement when the process parameters are optimally maintained [22]. The different factors and levels in turn are represented in Table 4.3.

The experimental runs obtained from Design of Experiments (DoE) by Box-Behnken Design method (BBD) in RSM, which is used for deciding the number and proportion of alumina in Al 6061-Al_2O_3 MMC composite specimens to be turned in a lathe are tabulated in Table 4.4.

TABLE 4.4
Experimental Runs Showing Different Combinations of Composite Specimens

Specimen No.	Composition of AJiO3 (wt.%)	Spindle Speed (rpm)	Depth of Cut (mm)
1	5	220	0.4
2	5	340	0.2
3	5	340	0.6
4	5	460	0.4
5	15	220	0.4
6	15	340	0.6
7	15	460	0.4
8	15	340	0.2
9	10	340	0.4
10	10	220	0.2
11	10	220	0.6
12	10	460	0.6
13	10	460	0.2
14	5	340	0.2
15	10	460	0.6
16	10	460	0.6
17	10	460	0.6

Optimization of Process Parameters

4.4 CONDUCTION OF TESTS AND OBTAINED RESULTS

To measure various physical as well as mechanical properties, different tests are conducted. The tests are classified into:

- Physical tests
- Mechanical tests

4.4.1 Physical Tests

Physical tests are done to identify the measurable values related to surface texture and integrity. They are microscopic in nature. As the composite specimens undergo turning operation in the lathe, it is imperative to ensure the quality and good workmanship of turning, through various physical tests. This part of the section deals with the detailed procedures of the surface roughness test and roundness test.

4.4.1.1 Surface Roughness Test

The surface roughness test procedure elaborates the ways and means of identifying the surface finish throughout the surface of the cylindrical composite specimens [23]. This test is done by a measuring device called a profilometer also known as surface roughness (SR) tester, shown in Figure 4.4. The probe stylus tip of the profilometer is gently moved over the composite workpiece's surface throughout the entire length from various positions. When the stylus tip comes in contact with the cylindrical

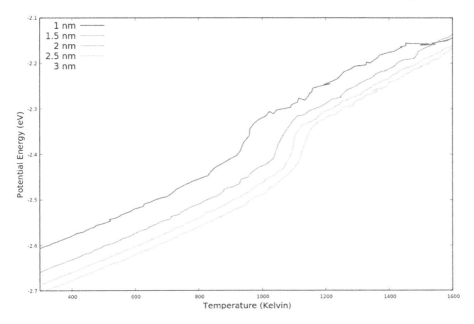

FIGURE 4.4 Profilometer.

TABLE 4.5
Experimental Test Results on Surface Roughness

Specimen No.	Surface Roughness (microns)
1	2.427
2	2.278
3	1.852
4	1.428
5	2.636
6	2.028
7	1.876
8	2.375
9	2.114
10	2.748
11	2.443
12	**1.365**
13	1.745
14	2.137
15	**1.365**
16	**1.365**
17	**1.365**

Bold values: Specimen(s) with smaller or lower values of surface roughness and roundness are desirable

surface, the surface's irregularities or waviness are measured and shown in the display panel. The least value of waviness indicates a good surface finish, which in turn is a result of proper turning.

Table 4.5 and Figure 4.5 provide the experimental test results on the 17 composite specimens to measure surface roughness.

From the results, it is evident that spindle speed and depth of cut play a crucial role in determining the surface finish of composite specimens. The specimens' surface roughness values with high spindle speed and depth of cut are far less compared to other specimens. Surface roughness increases with a reduction in both spindle speed and depth of cut. Particularly, the role of spindle speed is more significant in deciding the surface texture.

4.4.1.2 Roundness Test

The roundness test procedure tells how to measure the circularity maintained at the periphery throughout the entire composite workpiece [24]. It indicates how the cross-section of the workpiece resembles a circle. V-Block and Dial Gauge, shown in Figure 4.6 are used to measure a cylindrical job's Roundness. The specimen is kept in the V-Block, and the probe tip of the comparator dial gauge is made to pass through the surface for the entire length. The dial gauge values are taken at different points.

Optimization of Process Parameters

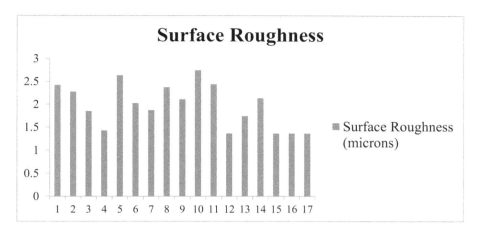

FIGURE 4.5 Graphical representation of surface roughness values.

FIGURE 4.6 Roundness measurement setup.

The value obtained from the diameter at one end of the specimen is considered as the reference value. When a particular location elsewhere indicates a different value, higher or lower than the reference value, the specimen is not round and said to be irregular. The % deviation from turned diameter of a specimen or simply % error is calculated using Eq. (4.1)

$$\% \text{Error} = (D \sim D')/L \tag{4.1}$$

where
D = reference value of diameter in mm.
D' = average of the enlarged or reduced diameters at various points in mm.
L = length of the workpiece in mm.
Lower the % error, the higher the roundness of the specimen.

Table 4.6 and Figure 4.7 provide the experimental test results conducted to calculate the % deviation from being round on the 17 composite specimens.

TABLE 4.6
Experimental Test Results on Roundness

Specimen No.	Average % Deviation from the Turned Diameter
1	7.6
2	6.6
3	4
4	3.3
5	8.43
6	5.33
7	5.3
8	7.26
9	5.66
10	8.76
11	8
12	**2.33**
13	3.6
14	6.33
15	**2.33**
16	**2.33**
17	**2.33**

Bold values: Specimen(s) with smaller or lower values of surface roughness and roundness are desirable

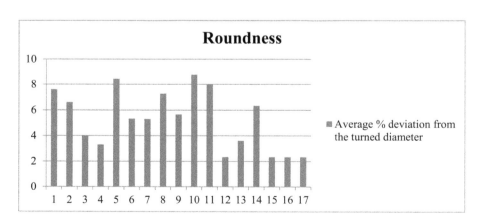

FIGURE 4.7 Graphical representation of roundness test results.

The results indicate that the combined effect of spindle speed and depth of cut is predominant in deciding the roundness of a composite specimen. Both spindle speed and depth of cut are inversely proportional to % deviation from the turned diameter. If they are more, roundness is high and vice versa.

4.4.2 MECHANICAL TESTS

Mechanical tests are done to identify the measurable values related to bulk material characteristics. They are macroscopic in nature. As the wrought aluminum alloy Al 6061 is melted and cast into composite specimens, it is mandatory to implicate that the casted components are defect-free. The binding strength between the matrix and reinforcements is intact [25]. This part of the section deals with the detailed test procedures to identify the specimens' hardness and impact strength.

4.4.2.1 Hardness Test

A hardness test is performed to evaluate the composite specimens' hardness. Hardness is the property of a material to remain resistant to external destructive loads in the form of abrasion, wear, and indentation [26]. For aluminum alloys and their associated composites, the Brinell Hardness test is normally performed. This test is done using Brinell Hardness tester as per ASTM Standard procedure for Brinell Hardness test 'E10-14'. Here, a static point load of 500 kgf is applied against the specimen with a help of a 10 mm diameter steel ball indenter for 10 seconds. The diameter of the impression created by the indenter's hard-pressing after the removal of load gives the hardness value. Equation (4.2) indicates the formula for calculating Brinell Hardness Number (BHN):

$$\text{BHN} = \frac{P}{\left(\frac{\pi D}{2}\right)\left(D\sqrt{(D^2 - d)^2}\right)} \tag{4.2}$$

where
P = load applied over indenter (500 kgf)
D = diameter of indenting steel ball (10 mm)
d = diameter of the impression created by ball indenter in mm.
Brinell hardness test results are represented in Table 4.7.

The pictorial representation in Figure 4.8 shows the graphical results of the Brinell hardness test.

From the above-mentioned results, we can clearly understand that the influence of alumina % is more when compared to spindle speed and depth of cut. As hardness is a mechanical property, the composition is more predominant than machining parameters. Higher composition gives more hardness. As far as the other two parameters are concerned, the depth of the cut has a slight edge over spindle speed.

4.4.2.2 Impact Test

An impact test is performed to measure the impact strength of the specimen. Impact strength is the property of a material to absorb maximum energy before being fractured by a sudden impact load with high magnitude and momentum. Izod Impact test is preferred for aluminum alloys and their associated composites over other test procedures. This test done in an Impact testing machine with samples from each composite specimen are prepared as per ASTM Standard procedure for impact test 'E23-18'. Here, a cylindrical sample having a length of 75 mm and diameter 11.4 mm

TABLE 4.7
Experimental Results for Hardness Test

Specimen No.	Brinell Hardness, BHN
1	43.38
2	37.98
3	39.88
4	40.44
5	44.03
6	47.26
7	**49.87**
8	46.74
9	42.45
10	41.05
11	42.43
12	45.37
13	43.96
14	37.98
15	45.37
16	45.37
17	45.37

Bold values: Specimen(s) with higher values of hardness and impact strength are desirable

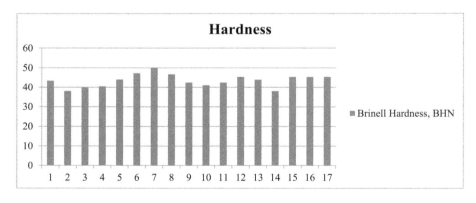

FIGURE 4.8 Hardness test results.

with a notch of 3.3 mm depth and notch divergence 45° created at 28 mm from one end is prepared. Then, the specimen is placed vertically inside the testing machine's anvil-like cantilever beam fixed at one end. Then, a hammer-like striker is released from a distant position to strike the sample just above the notch with a sudden impact, resulting in the sample's breakage. The striker follows an oscillating motion. The impact energy absorbed by the sample before deformation is recorded.

Table 4.8 and Figure 4.9 represent experimental impact test results.

TABLE 4.8
Experimental Test Results for Izod Impact Strength

Specimen No.	Izod Impact Strength (Nm)
1	2
2	2
3	2
4	4
5	6
6	**8**
7	**8**
8	6
9	4
10	4
11	4
12	4
13	4
14	2
15	4
16	4
17	4

Bold values: Specimen(s) with higher values of hardness and impact strength are desirable

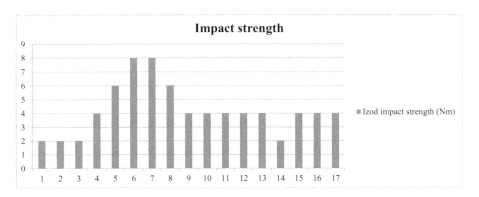

FIGURE 4.9 Bar graph showing impact test results for 17 composite specimens.

The above results infer that alumina composition is more significant in influencing the impact strength than the individual or combined effects of spindle speed and depth of cut. The composition of alumina is directly proportional to impact strength. If composition increases, impact strength also increases, and if it reduces, then impact strength also goes down.

4.5 MACHINE LEARNING INSPIRED OPTIMIZATION

The optimization technique discussed in this section of the chapter, Response Surface Methodology (RSM), is induced by Machine Learning (ML) [27]. It is one of the many software-driven optimization methods that rely heavily on identifying the optimal solution by predicting the R^2 value, also known as the goodness of fit value. This value should be very close to 1 so that the predicted properties have a reasonable agreement with experimental values through the conduction of tests [28].

The Historical Data Design (HDD), a category of RSM, uses imported data from other designs or techniques to predict apt values for properties, otherwise identified through conduction of tests [29,30]. This is exactly where HDD and Data Science converge.

The LRM is one of the algorithms used in ML for data prediction and application.

The Desirability indicates the closeness and accuracy of predicted values to experimental test results. It indicates the degree of correlation of dependent variables with independent variables [31]. Composition of Alumina, spindle speed, and depth of cut (process parameters) are independent variables, whereas surface roughness, roundness, hardness and impact strength (properties/criteria) are depending upon those independent variables.

The procedural steps for selecting the best possible composite specimen from a set of given alternatives using the LRM of HDD are described below:

Step 1: Preparation of Factors and Responses table
   The process parameters namely composition of alumina, spindle speed, and depth of cut, that led to the fabrication of composite specimens and turning respectively are considered as factors. The physical and mechanical properties obtained from experimental testing namely surface roughness, roundness, Brinell hardness, and Izod impact strength are collectively taken as responses/estimators. The 17 experimental runs, which indicate 17 different proportions of composite specimens are imported from Box-Behnken Design in RSM as shown in Table 4.9.

Step 2: Categorizing the attributes
   Table 4.10 represents the categorization of attributes/properties into minimization or maximization criteria.

Step 3: Creation of Analysis of Variance (ANOVA) table for all the four attributes/responses
   Table 4.11 indicates the ANOVA table for surface roughness.

Equation (4.3) indicates linear regression equation for surface roughness:

$$\text{Surface Roughness} = +3.79612$$

$$-0.014315 * \text{Composition of alumina}$$

$$-4.48845\text{E-}003 * \text{Spindle speed}$$

$$-0.55640 * \text{Depth of cut}$$

TABLE 4.9
Factors and Responses

Specimen No.	Factors			Responses			
	A: Composition of Alumina (wt.%)	B: Spindle (rpm)	C: Depth of Cut (mm)	Surface Roughness (microns)	Roundness (%)	Brinell Hardness (BHN)	Izod Impact Strength (Nm)
1	5	220	0.4	2.427	7.6	43.38	2
2	5	340	0.2	2.278	6.6	37.98	2
3	5	340	0.6	1.852	4	39.88	2
4	5	460	0.4	1.428	3.3	40.44	4
5	15	220	0.4	2.636	8.43	44.03	6
6	15	340	0.6	2.028	5.33	47.26	8
7	15	460	0.4	1.876	5.3	49.87	8
8	15	340	0.2	2.375	7.26	46.74	6
9	10	340	0.4	2.114	5.66	42.45	4
10	10	220	0.2	2.748	8.76	41.05	4
11	10	220	0.6	2.443	8	42.43	4
12	10	460	0.6	1.365	2.33	45.37	4
13	10	460	0.2	1.745	3.6	43.96	4
14	5	340	0.2	2.137	6.33	37.98	2
15	10	460	0.6	1.365	2.33	45.37	4
16	10	460	0.6	1.365	2.33	45.37	4
17	10	460	0.6	1.365	2.33	45.37	4

TABLE 4.10
Categorization of Attributes

S. No.	Attributes	Criteria/Type
1	Surface roughness	Minimize (lesser-the-better)
2	Roundness	Minimize (lesser-the-better)
3	Brinell hardness	Maximize (more-the-better)
4	Izod impact strength	Maximize (more-the-better)

$$+9.95833\text{E-}005 * \text{Composition of alumina} * \text{Spindle speed}$$

$$+0.011363 * \text{Composition of alumina} * \text{Depth of cut}$$

$$-1.43087\text{E-}003 * \text{Spindle speed} * \text{Depth of cut} \quad (4.3)$$

The predicted values of the four responses are obtained from substituting apt factor values as per the regression equation for all 17 composite specimens.

TABLE 4.11
ANOVA Table for SR

Source Indicator	Sum of Squares (SS)	DOF	Mean Sum of Squares (MSS)	F-Value	p – value Prob > F	Inferences
Model	3.578556199	6	0.596426033	133.1376966	<0.0001	Significant
A	0.126061836	1	0.126061836	28.14025813	0.0003	
B	2.207136596	1	2.207136596	492.6898995	<0.0001	Significantly influencing parameter
C	0.349143049	1	0.349143049	77.93774701	<0.0001	
AB	0.01428025	1	0.01428025	3.187720665	0.1045	
BC	0.006507268	1	0.006507268	1.452590341	0.2559	
AC	0.000613973	1	0.000613973	0.137054593	0.7189	
Residual(Res.)	0.044797683	10	0.004479768			
Lack of fit (LF)	0.034857183	6	0.005809531	2.337721647	0.2153	Non-significant
Pure error	0.0099405	4	0.002485125			
Cor. total	3.623353882	16				

Std. Dev.	0.066931071	R^2	0.987636404
Mean	1.973352941	Adj R^2	0.980218247
C.V. %	3.39174356	Pred R^2	0.952652079
PRESS	0.171558274	Adeq. precision	31.36969513

If experimental results > predicted values,

$$\% \text{Error} = ((\text{Experimental} - \text{predicted})/\text{Experimental}) * 100$$

If predicted values > experimental results

$$\% \text{Error} = ((\text{Predicted} - \text{experimental})/\text{Predicted}) * 100$$

Table 4.12 indicates the deviation of predicted values from experimental values in measuring surface roughness and vice versa.

Figure 4.10a and b indicates the 3D surface plot showing effects of various factors on surface roughness.

Table 4.13 indicates the ANOVA table for % Error in Roundness.

Equation (4.4) indicates Linear Regression Equation for Roundness:

Roundness = +0.14146

$$-1.17805E-003 * \text{Composition of alumina}$$

$$-2.07509E-004 * \text{Spindle speed}$$

$$-0.031557 * \text{Depth of cut}$$

$$+4.87500E-006 * \text{Composition of alumina} * \text{Spindle speed}$$

$$+1.78610E-003 * \text{Composition of alumina} * \text{Depth of cut}$$

TABLE 4.12
% Error in Predicted Values While Comparing Them with Experimental Values

Specimen No.	Experimental Values	Predicted Values	% Error
1	2.427	2.520983	3.7280299
2	2.278	2.17072	4.70939421
3	1.852	1.77627	4.08909287
4	1.428	1.426007	0.13956583
5	2.636	2.642405	0.24239282
6	2.028	2.039938	0.58521386
7	1.876	1.786469	4.77244136
8	2.375	2.388936	0.58335594
9	2.114	2.093966	0.94768212
10	2.748	2.733212	0.53813683
11	2.443	2.430176	0.52492837
12	1.365	1.386032	1.51742528
13	1.745	1.826444	4.4591567
14	2.137	2.17072	1.55340164
15	1.365	1.386032	1.51742528
16	1.365	1.386032	1.51742528
17	1.365	1.386032	1.51742528

$$-8.71235\text{E-}005 * \text{Spindle speed} * \text{Depth of cut} \tag{4.4}$$

Table 4.14 indicates the deviation of predicted values from experimental values in measuring % Error in roundness.

Figure 4.11a and b indicates the 3D surface plot showing effects of various factors on roundness.

Table 4.15 indicates the ANOVA table for Brinell Hardness.

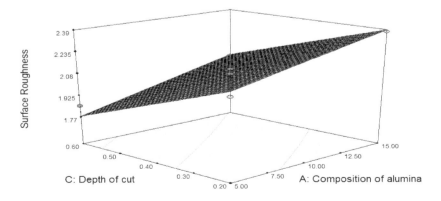

FIGURE 4.10 (a) 3D plot showing the influence on SR by factors A and C.

(*Continued*)

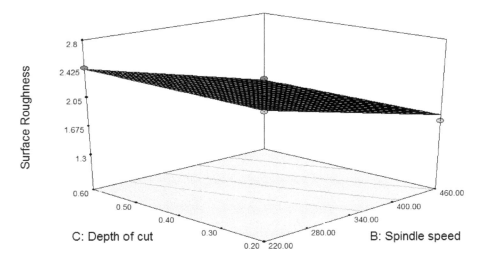

FIGURE 4.10 (CONTINUED) (b) 3D plot showing the influence on SR by factors B and C.

TABLE 4.13
ANOVA Table for Roundness

			2FI Model [Partial Sum of Squares – Type III]			
Source Indicator	Sum of Squares (SS)	DOF	Mean Sum of Squares (MSS)	F-Value	p – value Prob > F	Inferences
Model	81.94701841	6	13.6578364	46.04499699	<0.0001	Significant
A	3.096605849	1	3.096605849	10.43966283	0.0090	
B	50.06839315	1	50.06839315	168.796795	<0.0001	Predominantly influencing parameter
C	7.586804872	1	7.586804872	25.57758031	0.0005	
AB	0.342225	1	0.342225	1.15375149	0.3080	
BC	0.241251016	1	0.241251016	0.813335435	0.3883	
AC	0.151693735	1	0.151693735	0.51140879	0.4909	
Residual(Res.)	2.966193353	10	0.296619335			
Lack of fit (LF)	2.9297433533	6	0.488290559	53.58469782	0.0009	
Pure error	0.03645	4	0.0091125			
Cor. total	84.91321176	16				

Std. Dev.	0.544627703	R^2	0.965067941	
Mean	5.264117647	Adj R^2	0.944108705	
C.V. %	10.34603973	Pred R^2	0.850643583	
PRESS	12.68233306	Adeq. precision	18.2536159	

TABLE 4.14
% Error in Predicted Values While Comparing Them with Experimental Values

Specimen No.	Experimental Values	Predicted Values	% Error
1	7.60	7.8564832	3.26460572
2	6.60	6.2857702	4.76105758
3	4.00	4.1958106	4.66681218
4	3.30	2.6250976	20.4515879
5	8.43	8.4653832	0.41797517
6	5.33	5.7469106	7.25451689
7	5.30	4.4039976	16.9057057
8	7.26	7.1224702	1.89434986
9	5.66	5.8377404	3.04467804
10	8.76	8.8182166	0.66018564
11	8.00	7.5036498	6.2043775
12	2.33	2.4390714	4.47184121
13	3.60	4.5900238	21.5690341
14	6.33	6.2857702	0.69873302
15	2.33	2.4390714	4.47184121
16	2.33	2.4390714	4.47184121
17	2.33	2.4390714	4.47184121

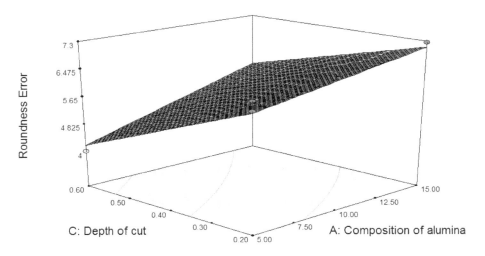

FIGURE 4.11 (a) 3D plot showing the influence on roundness error by factors A and C.

(*Continued*)

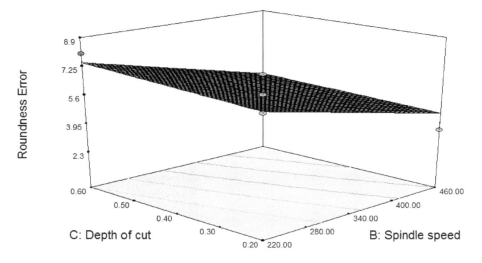

FIGURE 4.11 (CONTINUED) (b) 3D plot showing the influence on roundness error by factors B and C.

TABLE 4.15
ANOVA Table for Brinell Hardness

		2FI Model [Partial Sum of Squares – Type III]					
Source Indicator	Sum of Squares (SS)	DOF	Mean Sum of Squares (MSS)	F-Value	p – value Prob > F	Inferences
Model	158.6978693	6	26.44964489	20.9115892	<0.0001	Significant
A	98.7911098	1	98.7911098	78.10611876	<0.0001	Most significant parameter
B	11.46755258	1	11.46755258	9.066463827	0.0131	
C	5.893757294	1	5.893757294	4.65971592	0.0562	
AB	19.2721	1	19.2721	15.23688654	0.0029	
BC	0.010567912	1	0.010567912	0.008355191	0.9290	
AC	1.353001773	1	1.353001773	1.069708776	0.3254	
Residual (Res.)	12.6483189	10	1.26483189			
Lack of fit (LF)	12.6483189	6	2.10805315			
Pure error	0	4	0			
Cor. total	171.3461882	16				

Std. Dev.	1.124647451	R^2	0.926182665
Mean	43.46647059	Adj R^2	0.881892265
C.V. %	2.587390778	Pred R^2	0.692435597
PRESS	52.69998812	Adeq. precision	15.72360942

Optimization of Process Parameters

Equation (4.5) indicates Linear Regression Equation for Brinell Hardness:
Brinell Hardness = +42.53411

$$-0.35612 * \text{Composition of alumina}$$

$$-0.028047 * \text{Spindle speed}$$

$$+8.53221 * \text{Depth of cut}$$

$$+3.65833E\text{-}003 * \text{Composition of alumina} * \text{Spindle speed}$$

$$-0.53342 * \text{Composition of alumina} * \text{Depth of cut}$$

$$+1.82346E\text{-}003 * \text{Spindle speed} * \text{Depth of cut} \quad (4.5)$$

Table 4.16 indicates the deviation of predicted values from experimental values in measuring hardness.

Figure 4.12a and b indicates the 3D surface plot showing effects of various factors on the hardness.

Table 4.17 indicates the ANOVA table for Izod impact strength.

Equation (4.6) indicates Linear Regression Equation for Izod Impact strength:

$$\text{Izod Impact strength} = -1.68099$$

TABLE 4.16
% Error in Predicted Values While Comparing Them with Experimental Values

Specimen No.	Experimental Values	Predicted Values	% Error
1	43.38	41.1138415	5.22397077
2	37.98	38.7337083	1.94587173
3	39.88	41.3277428	3.50307745
4	40.44	38.9476096	3.6903817
5	44.03	43.4672875	1.27802071
6	47.26	47.0043448	0.54095463
7	49.87	50.0810476	0.42141219
8	46.74	46.5439903	0.41936183
9	42.45	43.4024466	2.19445362
10	41.05	41.5707302	1.25263674
11	42.43	43.0103987	1.34943813
12	45.37	45.321689	0.10648235
13	43.96	43.7069683	0.57559527
14	37.98	38.7337083	1.94587173
15	45.37	45.321689	0.10648235
16	45.37	45.321689	0.10648235
17	45.37	45.321689	0.10648235

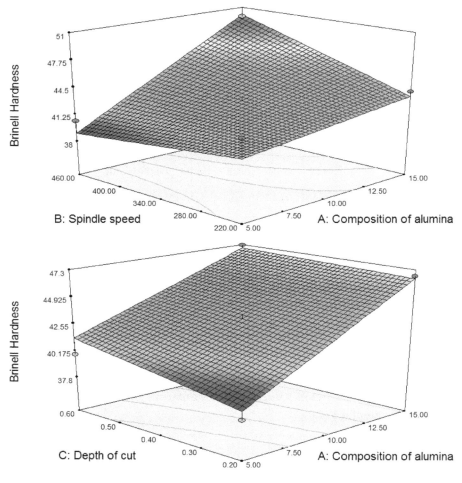

FIGURE 4.12 (a) 3D plot showing the influence on Brinell hardness by factors A and B. (b) 3D plot showing the influence on Brinell hardness by factors A and C.

$$+0.29172 * \text{Composition of alumina}$$

$$+8.38109\text{E}{-}003 * \text{Spindle speed}$$

$$+1.22659 * \text{Depth of cut}$$

$$+0.000000 * \text{Composition of alumina} * \text{Spindle speed}$$

$$+0.41655 * \text{Composition of alumina} * \text{Depth of cut}$$

$$-0.014048 * \text{Spindle speed} * \text{Depth of cut} \tag{4.6}$$

TABLE 4.17
ANOVA Table for Impact Strength

Source Indicator	Sum of Squares (SS)	DOF	Mean Sum of Squares (MSS)	F-Value	p-value Prob > F	Inferences
			2FI Model [Partial Sum of Squares – Type III]			
Model	48.04009275	6	8.006682125	11.40759259	0.0006	Significant
A	45.63963545	1	45.63963545	65.02548236	<0.0001	Most influencing parameter
B	1.018870517	1	1.018870517	1.451644961	0.2560	
C	0.15329324	1	0.15329324	0.218405926	0.6503	
AB	0	1	0	0	1.0000	
BC	0.627239139	1	0.627239139	0.893664622	0.3668	
AC	0.825071148	1	0.825071148	1.175527561	0.3037	
Residual (Res.)	7.01873078	10	0.701813078			
Lack of fit (LF)	7.01873078	6	1.169788463			
Pure error	0	4	0			
Cor. total	55.05882353	16				

Std. Dev.	0.837778657	R^2	0.872523052
C.V. %	19.78088496	Adj R^2	0.796036883
Mean	4.235294118	Pred R^2	0.502460018
Adeq. precision	10.14627753	PRESS	27.39396609

Table 4.18 indicates the deviation of predicted values from experimental values in measuring impact strength.

Figure 4.13a and b indicates the 3D surface plot showing effects of various factors on Izod impact strength.

Table 4.19 indicates the goals and limits of factors and responses.

The most optimized proportion of composite specimens is represented in Table 4.20.

4.6 CONCLUSION

The following inferences are drawn from the above-explained work:

- AMCs are used in a wide range of applications, belonging to different streams.
- Alumina is one of the most preferred ceramics to be reinforced inside aluminum alloy matrix, due to its attractive abrasive and refractory properties.
- Turning is one of the most important materials removal/machining processes, performed in a lathe to reduce the diameter.
- The composition of alumina is important to find out the mechanical properties of various composite specimens, whereas spindle speed and depth of cut improve their surface characteristics.

TABLE 4.18
% Error in Predicted Values While Comparing Them with Experimental Values

Specimen No.	Experimental Values	Predicted Values	% Error
	2	1.7089618	14.55191
2	2	2.3337846	14.3022882
3	2	1.7469926	12.65037
4	4	2.3718154	40.704615
5	6	6.2923618	4.64629672
6	8	7.1634926	10.4563425
7	8	6.9552154	13.0598075
8	6	6.0840846	1.38204193
9	4	4.3320886	7.66578504
10	4	3.5403558	11.491105
11	4	4.4609678	10.3333586
12	4	4.4495174	10.1026102
13	4	4.8775134	17.9909993
14	2	2.3337846	14.3022882
15	4	4.4495174	10.1026102
16	4	4.4495174	10.1026102
17	4	4.4495174	10.1026102

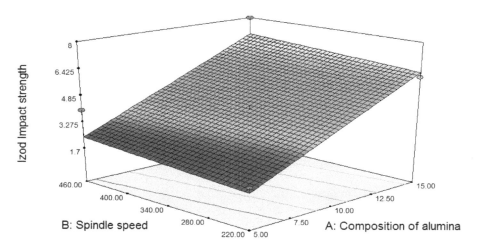

FIGURE 4.13 (a) 3D plot showing the influence on impact strength by factors A and B.

(*Continued*)

Optimization of Process Parameters

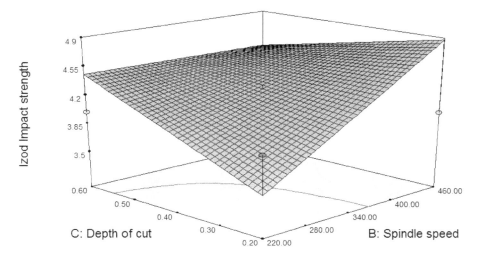

FIGURE 4.13 (CONTINUED) (b) 3D plot showing the influence on impact strength by factors B and C.

TABLE 4.19
Goals and Limits of Factors and Responses

Name	Goal	Lower Limit	Upper Limit
Composition of alumina	Is in range	5	15
Spindle speed	Is in range	220	460
Depth of cut	Is in range	0.2	0.6
Surface roughness	Minimize	1.365	2.748
Roundness error	Minimize	2.33	8.76
Brinell hardness	Maximize	37.98	49.87
Izod impact strength	Maximize	2	8

TABLE 4.20
Optimized Solution

Composition of Alumina	Spindle Speed	Depth of Cut	Surface Roughness	Roundness Error	Brinell Hardness	Izod Impact Strength	Desirability
15	460	0.6	1.57734997	3.5068695	50.35466	7.15770044	0.948070213

- Data science is an inter-disciplinary field more increasingly used in the fields of production and manufacturing to conduct computer-aided investigations and solve complex problems in industries.
- Machine Learning is an important sub-category in Data science that predicts values by training through historical data. RSM is a classical example.
- Multi-Objective Optimization is the process of selecting the best possible combination of the composite specimen from the given set of different specimen proportions/alternatives when different criteria are used with contrasting objectives.

REFERENCES

1. S. Dhinakaran & T.V. Moorthy, "Fabrication and characteristics of boroncarbide particulate reinforced aluminum metal matrix composites", *Anna University Newsletter 2014*, 81–88.
2. P. Kittali, J. Satheesh, G. Anil Kumar & T. Madhusudhan, "A review on effects of reinforcements on mechanical and tribological behaviour of Aluminium based Metal Matrix Composites", *International Research Journal of Engineering and Technology*, 3, No. 4 (2016), 2412–2416.
3. A. Mali & S. A. Sonawane, "Effect of hybrid reinforcement on mechanical behavior of aluminium matrix composite", *International Journal of Applied Engineering Research*, 11(2015), 130–142.
4. M.O. Bodunrin, K.K. Alaneme & L.H. Chown, "Aluminium matrix hybrid composites: A review of reinforcement philosophies; mechanical, corrosion and tribological characteristics", *Journal of Materials Research and Technology*, Elsevier, 169 (2015), 1–12.
5. D. Vengatesh & V. Chandramohan, "Aluminium alloy metal matrix composite: Survey paper", *International Journal of Engineering Research and General Science*, 2, No. 6 (2014), ISSN 2091-2730, 792–796.
6. A. Kusiak, "Smart manufacturing", *International Journal of Production Research*, 56, No. 1–2 (2018), 508–517, doi: 10.1080/00207543.2017.1351644.
7. R. Dubey et al., "The impact of big data on world-class sustainable manufacturing", *The International Journal of Advanced Manufacturing Technology*, 84 (2016), 631–645, doi: 10.1007/s00170-015-7674-1.
8. A.K. Choudhary, J.A. Harding & M.K. Tiwari, "Data mining in manufacturing: A review based on the kind of knowledge", *Journal of Intelligent Manufacturing*, 20, 501 (2009), 1–42, doi: 10.1007/s10845-008-0145-x.
9. T. Wuest, D. Weimer, C. Irgens & K.-D. Thoben, "Machine learning in manufacturing: Advantages, challenges, and applications", *Production & Manufacturing Research*, 4, No. 1 (2016), 23–45, doi: 10.1080/21693277.2016.1192517.
10. L. Monostori, A. Markus, H. Van Brussel & E. Westkampfer, "Machine learning approaches to manufacturing", *CIRP Annals*, 45, No. 2 (1996), 675–712.
11. I.D. Dinov, *"Data Science and Predictive Analytics"* Springer, Cham, 2018, doi: 10.1007/978-3-319-72347-1.
12. P. Sahoo, "Optimization of turning parameters for surface roughness using RSM And GA", *Advances in Production Engineering & Management*, 6, No. 3 (2011), 197–208.

13. S.A. Sajjadi, H.R. Ezatpour & M. Torabi Parizi, "Comparison of microstructure and mechanical properties of A356 aluminum alloy/Al_2O_3 composites fabricated by stir and compo-casting processes", *Materials & Design*, 34 (2012), 106–111.
14. V. Bharath, M. Nagaral, V. Auradi & S.A. Kori, "Preparation of Al6061-Al_2O_3 MMCs by stir casting and evaluation of mechanical and wear properties", *Procedia Materials Science*, 6 (2014), 1658–1667.
15. N. Valibeygloo, R. Azari Khosroshahi & R. Taherzadeh Mousavian, "Microstructural and mechanical properties of Al-4.5wt% Cu reinforced with alumina nano particles by stir casting method", *International Journal of Minerals, Metallurgy and Materials*, 49 (2013), 978–985.
16. B. Vijaya Ramnath, C. Elanchezhian, M. Jaivignesh, S. Rajesh, C. Parswajinan & A. Siddique Ahmed Ghias, "Evaluation of mechanical properties of aluminium alloy–alumina–boron carbide metal matrix composites", *Materials & Design*, 58, (2014), 332–338.
17. M.C. Gowri Shankar, P.K. Jayashree, R. Shetty & S.S. Achuthakini Sharma, "Individual and combined effect of reinforcements on stir cast aluminium metal matrix composites – A review", *International Journal of Current Engineering and Technology*, 3, No. 3 (2013), 922–934.
18. H.R. Ezatpour & M. Torabi, "Microstructure and mechanical properties of extruded Al/Al_2O_3 composites fabricated by stir-casting process", *Transactions of Nonferrous Metals Society of China*, 23, No. 5 (2013), 1262–1268.
19. W.T. Rashid & H. Zinalabiden, "Study the effect of stirring speed and particle size on the tensile strength of the aluminium matrix composites", *Diyala Journal of Engineering Sciences*, 9 (2012), 100–112.
20. N.G. Siddesh Kumar, G.S. Shiva Shankar, M.N. Ganesh & L.K. Vibudha, "Experimental investigations to study the cutting force and surface roughness during turning of aluminium metal matrix hybrid composites", *Materials Today: Proceedings*, 4, No. 9 (2017), 9371–9374.
21. S. Mathur & A. Barnawal, "Effect of process parameters of stir casting on metal matrix composite", *International Journals of Science and Research*, 2 (2013), 395–397.
22. P. Bansal & L. Upadhyay, "Effect of turning parameters on tool wear, surface roughness and metal removal rate of alumina reinforced aluminum composite", *Procedia Technology*, 23 (2016), 304–310.
23. P. Gudlur, A. Boczek, M. Radovic & A. Muliana, "On characterizing the mechanical properties of aluminum–alumina composites", *Materials Science and Engineering*, 590 (2014), 352–359.
24. R. Rufin Rasheed, K.J. Vishnu Deth & A. Mathew, "Investigation of heat treatment on Al 6061- Al_2O_3 composite & analysis of corrosion behavior", *International Journal of Engineering Trends and Technology*, 30 (2015), 162–167.
25. K. Madheswaran, S. Sugumar & B. Elamvazhudi, "Mechanical characterization of aluminium – boron carbide composites with influence of calcium carbide particles", *International Journal of Emerging Technology and Advanced Engineering*, 5, No. 7 (2015), 492–496.
26. A. Hascalik & U. Caydas, "Optimization of turning parameters for surface roughness and tool life based on the Taguchi method", *The International Journal of Advanced Manufacturing Technology*, 38, 896–903 (2008), doi: 10.1007/s00170-007-1147-0.
27. D.T. Pham & A.A. Afify, "Machine-learning techniques and their applications in manufacturing", *Proceedings of the Institution of Mechanical Engineers, Part B: Journal of Engineering Manufacture*, 219, No. 5 (2005), 395–412, doi: 10.1243/095440505X32274.

28. M. Mia & N.R. Dhar, "Response surface and neural network based predictive models of cutting temperature in hard turning", *Journal of Advanced Research*, 7, No. 6 (2016), 1035–1044, ISSN:2090-1232, doi: 10.1016/j.jare.2016.05.004.
29. P. Sharma, D. Khandiya & S. Sharma, "Dry sliding wear investigation of Al6082/Gr metal matrix composite by response surface methodology", *Journal of Materials Research and Technology*, 5, No. 1 (2016), 29–36.
30. C.H. Chan et al., "Analysis of face milling performance on Inconel 718 using FEM and historical data of RSM", *IOP Conference Series: Materials Science and Engineering*, 270 012038 (2017), 1–12.
31. J.A. Harding, M. Shahbaz, Srinivas & A. Kusiak, "Data mining in manufacturing: A review", *Journal of Manufacturing Science and Engineering: Transactions of the ASME*, 128 (2006), 969–976.

5 Spark Plasma-Induced Combustion Synthesis, Densification, and Characterization of Nanostructured Magnesium Silicide for Mid Temperature Energy Conversion Energy Harvesting Application

P. Vivekanandhan, R. Murugasami, and S. Kumaran

CONTENTS

5.1 Introduction 102
    5.1.1 Global Evolution in Renewable Energy Technologies 102
    5.1.2 Thermoelectrics Technology 103
    5.1.3 Thermoelectrics – Advancements and Technological Growth............ 103
5.2 Magnesium Silicide 105
    5.2.1 Crystallographic Structure and Band Gap Configuration 105
    5.2.2 Phase and Thermodynamical Behaviour of Mg_2Si 106
5.3 Spark Plasma Assisted Combustion Synthesis of Magnesium Silicide 107
5.4 Phase and Structural Characterisation of Mg-Si Milled Powders
    with Varying Bi Doping Concentration 110
    5 4.1 Characterization of Spark Plasma Assisted
           In-Situ Synthesized Mg_2Si_{1-x}– Bx Compound 112
    5.4.2 Analysis of Structure and Texture of Mg_2Si Compound 114
5.5 Thermoelectric Properties of the Mg_2Si Compound 117
5.6 Conclusions 121
References 121

DOI: 10.1201/9781003121954-5

5.1 INTRODUCTION

Energy is a vital factor to humans and a country's economic progression. Recent statistics from International Energy Agency (IEA) reveals that globally over 1.3 billion people are still without the access to electrical facilities. Remarkably, more than 95% of these people are either in African or developing Asian countries. Out of this, 84% of the population is in rural areas. In view to the Indian context, "Energy Statistics 2014" by the Central Statistics Office (CSO) of the Government of India reveals that we are still in the lag by 18% approximately towards the fulfillment in country's electricity consumption especially in rural areas (Energy statistics 2014, Govt. of India) [1].

On the other hand, the depletion of fossil fuels and their associated pollution from those thermal systems have become a threat to humankind. Also, it is estimated that the conversion of fuel intake to work efficiency of thermal systems is lesser than 40%. Whereas, remaining goes as waste heat [2]. It not only reflects in uneconomic condition due to higher overhead operation cost but also ends up with the emission of residuals thus causing against to the eco-friendliness of the living system. As it is expected that, in the forthcoming decade, the human population will be increased to several folds and the concurrent industrial revolution has made the entire globe on alert to meet these challenges. It is predicted that the world energy need is estimated to approx. 50 TW per year [3]. Hence, energy and pollution have become the global mission and several growing countries are coming together and several frameworks on policies are derived on the development of renewable and sustainable energy generation systems for future sustainable energy and environmental challenges.

5.1.1 GLOBAL EVOLUTION IN RENEWABLE ENERGY TECHNOLOGIES

The impact on the "energy and environment" has been approaching critical levels which sound alarming to the entire human community in the world to look for the prospective alternative source [4]. Considering the factual statistics, for the past couple of decades, the research on energy and environment are emerging ever-important and it has now become a global challenge. As the consequence, there could be seen significant and intensive global research exploring various eco-friendly energy conversion and storage technologies [5]. But, the economic and performance endurance of some of these technologies are foreseen to be unsatisfactory to eradicate the primary fossil conversion technologies. However, the global scientific fraternity research is still thriving to achieve higher efficiency and sustainability thereby it could be possible to bring these technologies into wider domestic and commercial implications. Thus, the existing energy and environmental issues could be resolved in tandem. Figure 5.1 shows the statistics recorded in 2017 which shows the highly encouraging trend on the strong evolution of energy technologies. The various renewable technologies together brought the energy generation contribution of 25% next to coal-based energy generation. This attainment is ever superior to the earlier years. It is noted that the long use fossil approach shares the contribution with 38%. Furthermore, the set record contribution by the renewables is well in command than other technologies such as gas, nuclear, and oil.

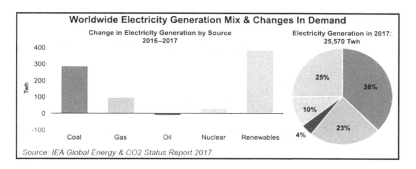

FIGURE 5.1 Worldwide electricity generation contribution (in percentage) [6].

5.1.2 THERMOELECTRICS TECHNOLOGY

Tremendous attempts are being made towards several alternative energy generation methods, however, the green mode of energy harvesting, and active refrigeration is well appreciated because of its no carbon footprints [7]. Though several renewable technologies are being appreciated, the direct and reliable conversion of waste heat into electricity with no carbon footprints has made thermoelectric technology to be attractive and promising. This is well convinced by its several positive traits such as simplicity in operation, low operating cost, noiseless operation, easy mobility, compactness, no fuel usage, free emission of residuals has put forth thermoelectric technology as a promising technology for future in various applications. Therefore, the attention towards the direct conversion of heat into electricity from thermoelectric generators (TEG) has become fascinating and spearheading in recent days. Besides, the efficiency of the thermal systems could be uplifted to 10%–15% more by means of cogeneration technologies by combining or integrating TEG with thermal systems. The energy conversion of thermoelectric materials is given by the unitless figure of merit $(ZT) = S^2\sigma/K.T$, Where, S is the Seebeck coefficient, σ and K are the electrical and thermal conductivity at absolute temperature T [8–10]. The inclination towards the thermoelectric being progressed considerably year to year owing to its proven advances and potential for economically viable and environmentally friendly energy conversion.

5.1.3 THERMOELECTRICS – ADVANCEMENTS AND TECHNOLOGICAL GROWTH

Thermoelectric generators (TEG) are composed of p-type and n-type thermoelectric couples possessing the ability to generate energy from the heat source. Hence, TEG could be employed for generating electric energy from the waste heat resources from thermal systems such as fuel cells, automobile exhaust systems, combustion chambers, industrial furnaces, heat treatment industries, sunlight, metal forming units, chemical processing industries, etc., as shown in Figure 5.2.

But there is a strong hindrance in achieving the higher efficiency of TEGs. From Figure 5.3, the efficiency of the TEG systems is far less than 10%. However, the spearheading growth and deeper scientific exploration in the past couple of decades in this field are commendably demonstrated by the introduction of nano-scale features in

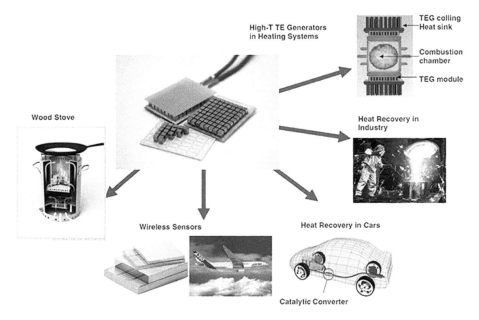

FIGURE 5.2 Thermoelectric technology in various applications [11].

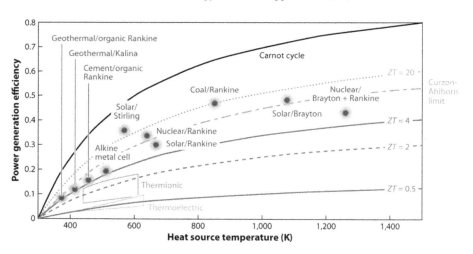

FIGURE 5.3 Comparison of power generation efficiency of various technologies with thermoelectric technology [14].

thermoelectric materials [11]. This leads to various breakthroughs in the development of novel materials. Besides, fundamental research is needed to be explored in advance with potential strategies for a better understanding of thermoelectric principles. Furthermore, the advanced research is also warranting the development of novel and superior nanostructured thermoelectric materials and processing methods to build an effective thermoelectric system [12,13].

5.2 MAGNESIUM SILICIDE

Late after the 1960s, the Magnesium-based compounds (i.e. Mg_2X, where $X = Si$, Ge, and Sn) and solid solutions have been realized as a potential candidate for thermoelectric applications. After that, considerable researches have been proved on these materials with notable electronic materials [15]. One among the compound is Magnesium silicide (Mg_2Si) which is possessing intrinsic semiconducting behavior with appreciable thermo-physical properties. The interest in the Mg_2Si compound is still a competitive system to explore enhanced thermoelectric properties by various engineering approaches. It is possessing a lower density and it could be made with cost-effective and abundantly available constituent elements, therefore, affirming the higher scale, sustainable, and low-cost devices [16]. The Laboratory for Physics of Thermoelements in Loffe Physical-Technical Institute has been synthesized Mg-based compounds of superior stoichiometry [17]. The interesting fact of these materials lies in the symmetrical crystal structure which favors a higher possibility of substitutions in the different crystallographic sites' consequences on the band engineering for better ZT. There are wide results available on Mg_2Si synthesis by liquid and solid-state synthesis approaches with notable properties.

5.2.1 Crystallographic Structure and Band Gap Configuration

Semiconducting silicides portray various kinds of crystalline structures. Notably, few silicides possessing identical stoichiometry with distinct crystalline structures with non-semiconducting phases. Among that, Mg_2Si is a semiconductor with an antifluorite crystal structure that shows n-type conductivity and is known as a substance with strong thermoelectric properties. The electron density distribution shows that a large peak exists at the center site of the tetrahedron enclosed by four Mg^{2+} ions at the lowest end of the conduction band. In other words, the researchers learned that electron carriers generated by the loss of Silicon (Si) occupy this interstitial site. This type of semiconductor substance was previously unknown and is thought to be connected to the origin of thermoelectric properties.

Magnesium silicide possesses a face-centered-cubic (FCC) structure of CaF_2 type (Space group Fm3m) as shown in Figure 5.4a. The primitive translation vectors of Mg_2Si lies on the Bravais lattice at $a_1 = a(0, 1/2, 1/2)$, $a_2 = a(1/2, 0, 1/2)$, $a_3 = a(1/2, 1/2, 0)$, where a represents the lattice parameter. The lattice constant of the Mg_2Si is 0.6351 nm. In accordance with each silicon atom, there are eight Mg atoms present at $a(\pm 1/4, \pm 1/4, \pm 1/4)$ in the unit cell. The interatomic distance between the closest Si sites is 0.2750 runs. At 25×10^5 Pa and above 900°C the FCC of Mg_2Si is transformed into hexagonal with lattice parameters $a = 0.720$ run and $c = 0.812$ nm [18,19]).

It has been theoretically estimated that the indirect gap value of Mg_2Si is 0.78 eV. Minimum bandgap is witnessed at the X point whereas the upper valence band shows the density of state (partial) confirm that Si 3p electrons are highly dominating with the synergistic association of Mg 3s electrons, wherein the lower conduction band possesses Mg 3p electrons that dominate with the participation of Si 3s electrons. The conduction band (lowest) states correspond to the non-bonding Mg orbitals (Figure 5.4b and c) thereby attributing to having the intrinsic and n-type semiconducting

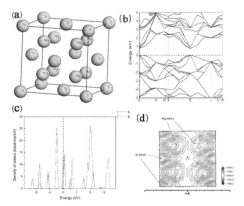

FIGURE 5.4 (a) Crystal structure of Mg_2Si i, (b) band structure, (c) density of states and gap, and (d) charge carrier mapping [20].

nature. Figure 5.4c shows the valence electron charge density difference map in (101) of different color patterns attributed by Mg and Si (Figure 5.4d). The plot infers that the loss of electrons is represented in black color and electron enrichment is indicated in gray color. The white color represents the regions that are having a lesser charge in the electron density. At the Si atom site, the remarkable charge transfer from Mg to Si disrupts its charge anisotropy. The combined nature of bonds (i.e. ionic and covalent bonds) is showcased by the difference between the crystal density and a sum of superimposed neutral spherical atoms. The density contour region represents the covalent characteristics (i.e. primary bond directions). On the other hand, the Valence electron charge density shows a higher concentration near the Si atom sites. The Si sites and Mg sites make the sp3 hybrids with tetrahedral kind of bonding.

5.2.2 Phase and Thermodynamical Behaviour of Mg_2Si

The binary phase diagram of Mg-Si is portrayed in Figure 5.5. It confirms the Mg_2Si is the sole and primary stoichiometric silicide in this system. It is understood that it is showing the congruent melting at temperature 1085°C. There are two eutectics on the sides of the compound. The 1.16 % of silicon is at the lowest side at 637.6°C and 53 % of silicon observed at 945.6°C. The formation enthalpy of the Mg_2Si compound is −80.02 kJ/mol and $Cp = 73.353 + 14.989 \times 10^{-3}$ in the temperature range of 298–873 K [21].

At normal pressure, the compound is stable and that possesses an orthorhombic structure and it showcases an intrinsic semiconducting nature. On the other hand, it could form a compound of metastable behavior that forms cubic and trigonal crystal structures. They can be produced by the application of pressure up to 10^9 Pa and a temperature of around 1000°C. The metastable phase phases can transform to orthorhombic at ambient pressure. The transformation of trigonal to orthorhombic transformation occurs in the temperature range of 400°C–420°C. However, the change in crystal structure from cubic to orthorhombic transformation requires thermal processing at 490°C. The respective formation enthalpies are −4.6 and −3.3 kJ/mol.

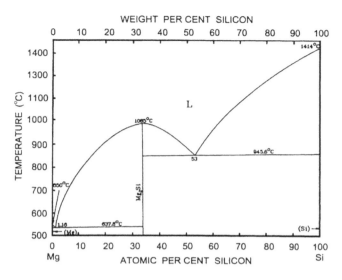

FIGURE 5.5 Phase diagram of Mg$_2$Si system [22].

5.3 SPARK PLASMA ASSISTED COMBUSTION SYNTHESIS OF MAGNESIUM SILICIDE

The spark plasma sintering is a proven method to synthesize a wide range of materials including ceramics, semiconductors, alloys, composites, and functionally graded materials method in a rapid fashion due to its superior heating rare more than 800°C per minute bypassing the high-density pulse current through the graphite die filled with powder. In this current approach, the spark plasma-assisted combustion synthesis and simultaneous synthesis of Mg$_2$Si with substantial Bi doping are carried out. The schematic representation is presented in Figure 5.6.

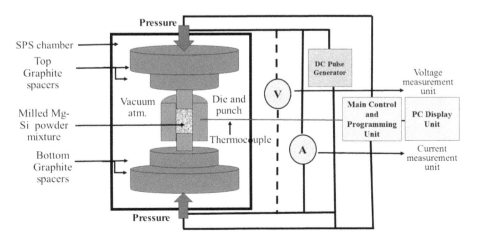

FIGURE 5.6 Systematic representation of die assembly arrangement in the SPS.

In the current synthesis approach, the three-step sintering cycle (RT to 700°C) under the constant pressure (50 MPa) plays a prominent role in synthesizing the Mg_2Si compound with Bi doping and simultaneous densification of the milled Mg-Si mixture. While SPS, the pulse current (I) passes through the die filled with Mg-Si mixture, and accordingly, it exhibits certain resistance (R) thus the Ohmic nature is observed. As a consequence of the pulse current, the temperature is raised in the Mg-Si powder mixture resulting in Joules heating. The power (P) contribution to cause Joule heating in the powder mixture is proportional to the product of the square of the current (I) and resistance (R) imposed (i.e. $P \alpha I^2 R$) [23].

Figure 5.7a and b portrays the P and R characteristics of the Mg-Si powder mixture with the increase in SPS temperature. It is understood that the P-value of the Mg-Si mixture gradually increases with the increase in temperature. Furthermore, it reveals that the stoichiometric inhomogeneity of Mg-Si powder (i.e. increasing of Bi doping concentration) is significantly complementing the higher power consumption. Notably, the R characteristic is observed attributing to lower resistance in the Mg-Si mixture against the current flow path. The instantaneous densification rate and the corresponding relative density profiles of Mg-Si powder mixture during SPS from RT to 700°C (Figure 5.7c and d). The instantaneous densification rate and relative density profile are manifested in three distinct stages (i.e. I, II and III) distinguished as (i) primary (RT-150°C) (ii) intermediate (150°C–400°C) and (iii) final densification (400°C–700°C) stages.

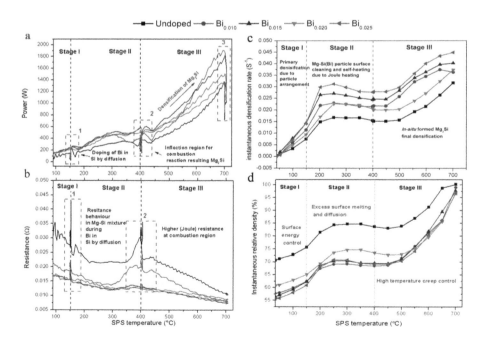

FIGURE 5.7 Characteristics of milled Mg-Si powder mixture with varying Bi concentration as a function of SPS temperature: (a) power, (b) resistance, (c) instantaneous densification rate, and (d) relative density [29].

In the primary regime (>150°C), the lower densification rate is observed as the milled Mg-Si powder undergoes self-arrangement. It is due to the applied uniaxial pressure resulting in the reduction of initial porosity in the Mg-Si green compact. At this regime, initially, the temperature is expected to be lower to cause remarkable thermodynamical phenomena in the Mg-Si mixture. But, it helps to improve the surface relationship between the neighbor particles, thus creating a better contact network for uniform electro-migration generated by the high-density DC pulse current. An increase in the temperature beyond 150°C results in higher electron density, thus creating spark discharge and causes plasma generation at particle surface contact vicinities. Also, the plasma encapsulates the Mg-Si powder particles resulting in self-healing.

Notably, increasing pulse current results in the higher electrical discharge momentum and pulse pressure across particle capacitor gaps hence breaks the thin inherent oxide layer formation in the Mg-Si mixture. The self-heated particle surfaces start to melt and bridge with the neighborhood particles by necking phenomena leading towards surface diffusion. The thin melted surface of low-temperature elements in the powder mixture (Bi and Mg) results in the formation of a thin amorphous layer which helps to glide the particles. Further increase in the temperature, volume diffusion is initiated and resulted in grain diffusion thus the particle grain boundaries are getting vanished. The melt viscosity (η) of the amorphous layer and the volume diffusion coefficient (D_v) of the diffusion species can be given by the Stocks-Einstein relation Eq. 5.1 [24]:

$$\eta = \frac{kT}{3\pi r D_V} \tag{5.1}$$

Where, k is Boltzmann's constant, t is the temperature, and r is the ionic radius. The Mg-Si mixture strain point viscosity is calculated as ~180°C. The theoretically calculated temperature provides a close concurrence with the current experimentation (i.e. 150°C) where the densification is observed higher. This is due to the atomic diffusion at the surface level, thereby reduces the micro-pores (internal voids) in the interparticle junctions and improves the densification. It can be understood by the second bump in the densification curves marked as an intermediate densification stage. The Bi doping with varying concentrations with Mg_2Si is significantly attributed to the densification kinetics. This can be witnessed from the higher densification rate of the Mg-Si mixture with the higher Bi doping concentration. It promotes a higher in-situ doping of the Bi diffusion. To overcome the loss of Bi with the increase in sintering temperature due to its low melting temperature (271°C), the 10 minutes soaking is provided at 150°C. The soaking period and intimate mixing of Bi in Mg-Si restrict the loss of Bi from sudden vaporization. At this regime, there could be seen the increase in R and reducing trend in the P (marked as 1 in Figure 5.7a and b). Increasing the temperature range to 375°C–400°C indicates the combustion reaction of Mg (fuel) with Si (reactant). The spark plasma-assisted combustion reaction resulting Mg_2Si is represented as follows:

$$Mg(\text{fuel}) + Si(\text{reactant})(\Delta SPS) \rightarrow Mg_2Si \tag{5.2}$$

The inflection temperature to stimulate the combustion reaction in a 10 h milled Mg-Si mixture is close to 375°C and the standard formation to form Mg_2Si is -77.8 kJ/mol (-25.9 kJ/g-atom) [18]. It is marked as a window in Figure 5.7b. The exothermic combustion results in higher heat energy, thus a sudden rise in the temperature. It seems to be a sudden dip in the P profiles, and substantially the R-value sensibly increases (marked as 2 in Figure 5.7). The combustion reaction is progressed until the complete consumption of Mg and Si. Combustion reaction results in the Mg_2Si phase formation. The chemical potential of m_{Mg} and m_{Si} to form m_{Mg2Si} (bulk) and the substantial substitution of Bi atoms occurs at Si sites that concedes from the density functional theory model by J. Tani et al. [25] are given in Eqs. (5.3) and (5.4), respectively.

$$m_{Mg} + m_{Si} = m_{Mg2Si} \tag{5.3}$$

$$Mg_{32}Si_{16} + Bi \rightarrow Mg_{32}Si_{15}Bi + Si \tag{5.4}$$

At this stage, no remarkable observation in densification rate is observed. However, at this point, the in-situ formed Mg_2Si undergoes creep due to the assistance of imposed pressure at elevated temperatures. It is because of Mg-Si powder sinter forging phenomena and the densification occurring by power-law creep. The deformation rate (ε) of the Mg-Si powder under stress (σ) is expressed by the Norton law [26]:

$$\varepsilon = \frac{d\varepsilon}{dt} = A_0 \exp\left(\frac{Q}{RT}\right)\sigma^n \tag{5.5}$$

where $A0$ is a material constant, Q is activation energy and n is a stress exponent. The increase in the newer surface area by nanostructure phenomena by mechanical milling enhances the reactive enthalpy at lower activation energy [27]. The density value is observed for Mg_2Si is 85%, and it is noticed decreasing with the substantial increase in Bi concentration. The lowest density of 67% is observed for Mg_2Si with Bi 0.025 at.%. Notably, there could be observed a significant increase in the bulk relative density by increasing the SPS temperature up to 700°C followed by soaking for 10 minutes. The Mg_2Si with no doping exhibits a superior density (100%) that close to theoretical density. The relative density values of Mg_2Si with increasing Bi concentration (>96%).

5.4 PHASE AND STRUCTURAL CHARACTERISATION OF MG-SI MILLED POWDERS WITH VARYING BI DOPING CONCENTRATION

XRD patterns of milled Mg-Si powder mixture with Bi of varying atomic concentration (0.010–0.025 at.%) are shown in Figure 5.8. The Mg-Si combustible powder mixture clearly shows the peaks representing its parental elements (i.e. Mg and Si). Besides, the trace of the Mg_2Si phase is observed. The crystallite size and lattice strain before and after 10 h milling is summarized in Table 5.1. The average crystallite size of Mg is significantly reduced to 47 nm, wherein the crystallite size of Si is increased

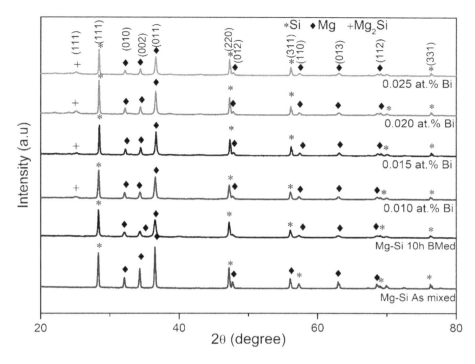

FIGURE 5.8 XRD patterns of 10 h milled Mg-Si powder mixture with Bi of varying atomic concentration (0.010–0.025 at.%).

TABLE 5.1
Summary of Crystallographic Details of Milled Mg-Si Powder and Sintered Mg_2Si Doped with Varying Bi Concentration [28]

Sample Code	Bi Doping (at. %)	Milled Powder (10 hours)				Spark Plasma Sintered		
		Avg.cry. Size (nm)		Lattice Strain		Avg.cry. Size (nm)	Lattice Strain	Inter Planar Distance (d)(Å)
		Mg	Si	Mg	Si	Bulk Mg_2Si		
As mixed	–	47.0	51.8	0.235	0.273	–	–	–
B0	–	24.8	48.2	0.447	0.294	44 ± 0.12	0.231 ± 0.010	2.2398
B1	0.010	31.2	43.1	0.354	0.328	47 ± 0.19	0.215 ± 0.012	2.2580
B2	0.015	35.7	56.9	0.310	0.248	46.2 ± 0.16	0.220 ± 0.011	2.2561
B3	0.020	35.5	57.7	0.311	0.244	45.7 ± 0.19	0.222 ± 0.012	2.2565
B4	0.025	34.4	61.2	0.320	0.231	43.4 ± 0.12	0.234 ± 0.010	2.2569

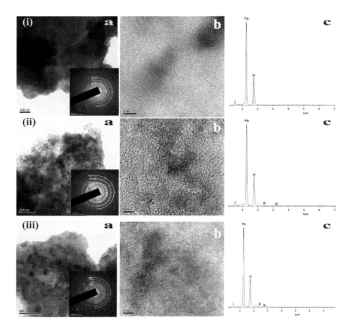

FIGURE 5.9 TEM images of 10 h milled powders (i) Mg-Si, (ii) Mg-Si powder with Bi 0.015 at. %) and (iii) Mg-Si powder with Bi 0.025 at. % with the SAED pattern as an inset ((a) TEM image of powders, (b) HRTEM image, and (c) EDS spectrum).

to 51.8 nm. During mechanical milling, the ductile Mg and brittle Si powder undergo severe plastic deformation and fragmentation. Notably, the average crystallite size of Mg and Si is varying with increase in the Bi doping concentration. However, the variation in the average crystallite size and the corresponding lattice strain is negligible.

The TEM observation of milled undoped Mg-Si powder mixture and with Bi doping of 0.010 and 0.025 concentration at.% is shown in Figure 5.9. It reveals that the mechanical milling results in the intimate mixing and finer nano Mg-Si particles. The corresponding selected area electron diffraction (SAED) ring-type patterns show the finer spots corresponding to the crystallographic planes of Mg and Si (shown as an inset in Figure 5.9ia). It confirms the polycrystalline Mg and Si particles of the nanoscale. The HRTEM image of intimately mixed Mg and Si powder is shown in Figure 5.9ib–iiib. It shows the Si particles (darker region) and Mg particles (lighter regions) in the nanocrystalline scale. The corresponding EDS patterns attest to the presence of Mg, Si, and Bi (Figure 5.9ic–iiic).

5 4.1 Characterization of Spark Plasma Assisted In-Situ Synthesized Mg_2Si_{1-x}– Bx Compound

The XRD patterns confirm the evolution of the Mg_2Si compound by in-situ combustion synthesis by SPS (Figure 5.10a). It is evidentially understood from the diffracted planes that representing the Mg_2Si phase in the XRD observation is well compliant

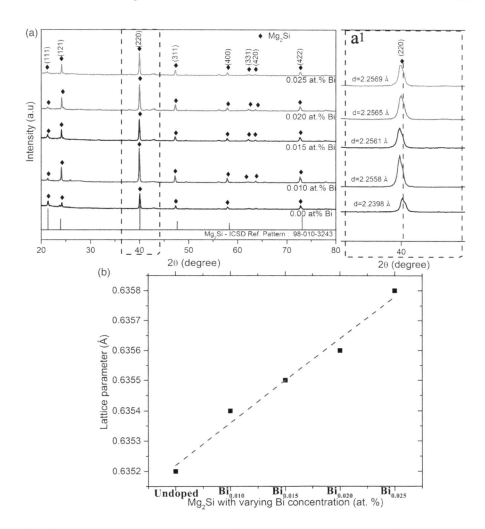

FIGURE 5.10 Spark plasma sintered Mg-Si powder mixture with varying Bi concentration: (a) XRD patterns (a1 – Shift in the principal plane of Mg_2Si phase (220)) and (b) change in lattice parameter of Mg_2Si [28,29].

to the standard ICSD source (Ref. 98-010-3243) of the Mg_2Si compound with antifluorite structure of the Fm3m group. The average crystallite size and lattice strain of Mg_2Si with varying Bi concentration are summarized in Table 5.1.

It can be seen that the trend of average crystallite size increases and with the decrease in the lattice strain. The Mg_2Si doped with Bi 0.010 at.% possesses a higher crystallite size of 47±0.19 nm with a lattice strain of 0.215±0.012. The undoped Mg_2Si show the crystallite size of 43.6±0.12 nm with a strain of 0.231 ±0.010%. Alongside, Bi doping in Mg_2Si significantly attributes the lattice defects that attribute the shift of principal plane (220) of Mg_2Si from its Bragg's position (2Θ). It is due to the substitution of Bi in Si sites of Mg_2Si cubic structure causing the lattice strain.

Figure 5.10b shows an increase in the lattice parameter (*a*) from 0.6352 Å (undoped Mg$_2$Si) to 0.6358 Å (Mg$_2$Si with Bi 0.025 at.%). Similarly, the significant change is attributed to the interplanar distance (*d*) of the Mg$_2$Si unit cell (Table 5.1). The increase in the lattice parameter and *d* spacing values of Mg$_2$Si$_{(1-x)}$-Bi_x is due to the higher atomic radius of Bi (230 pm) substituting Si (117.6 pm) of the lesser radius.

5.4.2 Analysis of Structure and Texture of Mg$_2$Si Compound

Figure 5.11a depicts the HRTEM image and the respective SAED pattern as an inset of sintered Mg$_2$Si doped with Bi 0.025 at.%. It confirms the polycrystalline Mg$_2$Si compound with the nanostructure features. The EDS spectrum (Figure 5.11b) confirms the parental constituent elements Mg and Si. The elemental mapping of constituent elements (Mg and Si) and dopant Bi$_{0.025}$ at.% is shown in Figure 5.11c. It portrays the chemical homogeneity of Mg and Si elements and the uniform distribution of Bi doping. Furthermore, the experimental quantification showed a closer agreement with the theoretically estimated stoichiometry.

The end properties of the materials are significantly manifested by its nature of the grain size its orientation, crystallographic structure, and phase. Furthermore, it is strongly emphasized that the superior properties are ascertained as superior at their certain orientation. Likewise, thermoelectric transport properties are also in compliance and dependent on the texture and grain orientation. The texture of the combustion synthesized Mg$_2$Si compound firmly lies on a ball milled Mg-Si intimate mixture which manifests the degree of mixing, particle size, and distribution. On the other part, the SPS processing parameters such as temperature and imposed pressure play a crucial role in contributing to several factors including flow and rearrangement of the nanocrystalline particles due to its surface area, surface tension interparticle friction

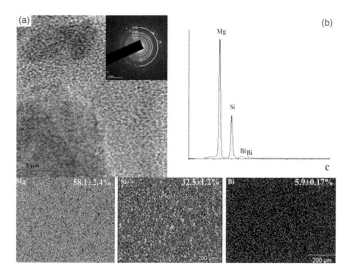

FIGURE 5.11 Sintered Mg$_2$Si compound doped with Bi$_{0.025}$ at.%: (a) HRTEM image (Inset: SAED pattern), (b) EDS spectrum, and (c) area elemental mapping [29].

in the confined die volume during the in-situ synthesis and simultaneous densification. This phenomenon yields the key attribute towards the sintered polycrystalline Mg_2Si with an imperfect orientation leading towards the anisotropy texture. The ball milling results in the revolution of newer and higher surface area due to nanostructuring that substantially enhances the reactive enthalpy during spark plasma-assisted combustion synthesis. Increasing SPS temperature from RT to 700°C induces thermal kinetics resulting from the grain boundary diffusion resulting in the immigration of atoms and formation of newer (lower and higher angle) grain boundaries, etc. Furthermore, the associated subgrain evolution, which is observed an intricately unstable and the need of thermal activation is higher with the further increase in temperature. It features the significances such as abnormal growth of individual subgrains, coalesces or bugles out, and forming new strain-free recrystallization nuclei. The free energy is highly influential in forming the critical nuclei size (ΔG_c) in mixed chemistry by an electrical field. It is understood from the following correlation [30]

$$\Delta G_c = \frac{16\pi\sigma^3}{3\left[|\Delta G_V| + \frac{1}{2}E^2(\varepsilon_2 - \varepsilon_1)\right]^2} \qquad (5.6)$$

where σ is the interfacial energy, ΔG_V is the driving free energy, E is the electrical field, and ε_1 and ε_2 are the die-electric constants of the major phases (i.e. Mg and Si, respectively). In the case of $\varepsilon_2 > \varepsilon_1$, ΔG_C is significantly reduced, and thus the nuclei formation or phase separation is induced by the applied pulse electric field. On the other hand, if $\varepsilon_1 > \varepsilon_2$, ΔG_C, the formation of nuclei is inhibited. Reaching the temperature ~375°C and above the uniform reactive momentum takes between the elemental phases (Mg and Si), while combustion reaction exhibits thermodynamically stable and homogenous Mg_2Si phase with further favoring Bi doping and simultaneously densification.

Figure 5.12 shows the (i) pattern quality map, (ii) inverse pole figure map, (iii) misorientation angle, and (iv) area fractions of multiple reference textures of dry-etched antifluoride (cubic) Mg_2Si. This random grain orientation in the recorded planes of (001), (101), and (111) planes of the cubic unit cell. It confirms that the sintered nano polycrystalline Mg_2Si possesses strong anisotropy behavior. The pattern quality maps reveal the fine equiaxed Mg_2Si compound. In addition to in-situ phase evolution, the doping of Bi in Mg_2Si has a substantial influence during the sintering kinetics. The Mg_2Si doped with Bi possesses the grains slightly coarser than the undoped. The average grain size of samples undoped and B doping (0.010 at.% and 0.025 at.%) are 3.5, 4.5, and 4.2 µm, respectively. The band contrast image based on its diffraction quality index is shown as superimposed colors corresponding to the individual grains. The weaker diffraction quality is observed from the grains oriented in the other planes than the (001), (101), and (111). It is due to the possibility of lattice rotation due to the existence of geometrically necessary dislocations [30]. Also, the grain boundaries in the EBSD grain orientation map seem to be black regions which reveal the data points of confidence index (CI) greater than 0.1. On the other hand, the weaker diffraction may also come due to certain crystallographic factors such as the neighboring planes overlapping, chemical inclusions, porosity, or deformation [31].

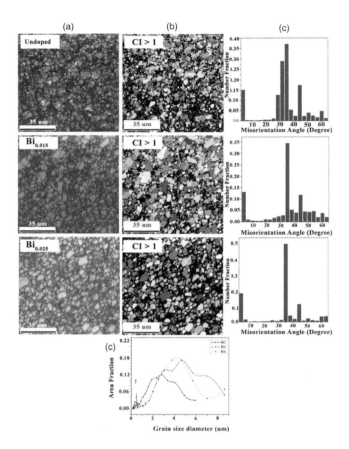

FIGURE 5.12 EBSD grain orientation map of Mg_2Si (undoped and doped with $Bi_{0.015}$ at.% and $Bi_{0.025}$ at.%: (a) pattern quality map, (b) crystal orientation map (after CI standardization greater than 0.1), (c) histogram showing the distribution of misorientation angle of grains, and (d) area fractions of multiple reference textures for Mg_2Si.

The area fraction concerning grain size diameter reveals that Mg_2Si without doping is narrow size (dia) grains of smaller area fraction (Figure 5.12d). It confirms that the Mg_2Si compound is thermodynamically stable and offers minimum growth during the sintering process. The grain misorientation plot of Mg_2Si doped with Bi in the cubic planes (001), (101), and (111) is shown in Figure 5.12d. The misorientation angle is varying from 0° to 80°. In order to understand more clearly, the misorientation is separated into low-angle grain boundaries (LAGBs, 2°–15°) and high-angle grain boundaries (HAGBs > 15°). As portrayed, the Mg_2Si grain misorientation angle distribution is dominated by HAGBs.

Figure 5.13a and b exhibits the pole figures and the corresponding stereographic triangle inverse pole figure (IPF) of bulk Mg_2Si. The center of the pole figure is the axis of the bulk Mg_2Si pellet. The normal direction (ND) in the {011} plane and the transverse direction (TD) in the {101} plane of pellet are denoted as *A*1 and *A*2, respectively. The substitution of Bi as a doping element, at a given temperature and constant pressure,

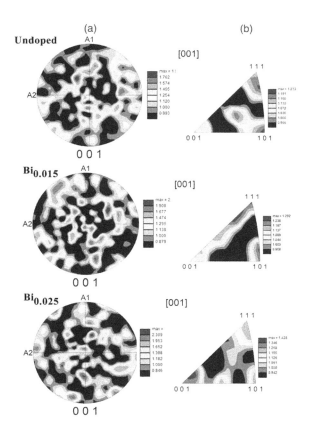

FIGURE 5.13 Crystallographic texture evolution of bulk nanocrystalline Mg_2Si (undoped and doped with $Bi_{0.015}$ at.% and $Bi_{0.025}$ at.%): (a) pole figures and (b) stereographic triangle inverse pole figure.

occupies the substitutional sites of Si in the antifluorite cubic structure and causes significant evolution in the polycrystalline texture of sintered Mg_2Si. It is evidentially witnessed by the variation from the basal pole density, which is varying random intensity denoted by the color contours that is corresponding to the texture evolved in the Mg_2Si. The maximum intensity is much higher than random for the sintered Mg_2Si. Where in, the sample Bi 0.025 at.% exhibits the superior intensity in the plane (101) than (111). Thus, it can be understood that the intensity of texture is increased with increasing doping of Bi. The maximum and minimum intensities of undoped Mg_2Si i are 1.974 and 0.893, respectively. Whereas, this case is 2.730 and 0.846 for $Bi_{0.025}$ at.%, respectively.

5.5 THERMOELECTRIC PROPERTIES OF THE MG_2SI COMPOUND

The temperature-dependent thermoelectric transport properties such as absolute Seebeck coefficient (S), electrical conductivity, thermal conductivity (K) are key parameters to estimate the figure of merit (ZT) of Mg_2Si doped with Bi at varying concentration. Figure 5.14 shows the S plot of the Mg_2Si doped with Bi. The S

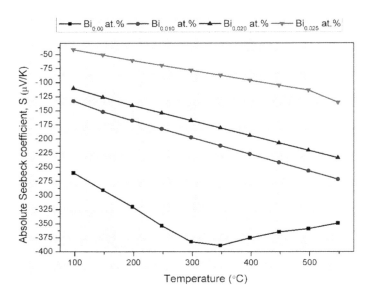

FIGURE 5.14 Absolute Seebeck coefficient of $Mg_2Si_{1-x}Bi_x$ compound [28].

values of Mg_2Si are recorded in the negative sign thus attests to the n-type semiconducting phenomena. Further doping of Bi in the Mg_2Si is strongly contributed by electrons as the major charge carriers. This phenomenon is mathematically expressed by [32]

$$S = -\frac{k}{e}\left(s - c\ln n\right) \tag{5.7}$$

where k denotes the Boltzmann constant, e is the electron charge, S is an energy-dependent scattering expression and c is a constant. This experimental observation is in good compliance with the theoretical modeling approach made by Nieroda et al. [33], in which it is emphasized that the doping of lower atomic concentration of Bi in Mg_2Si changes the conduction behavior as electron one (i.e. Fermi level shifts to conduction band). Where in, for the undoped Mg_2Si, the states are located close to the top of the valence band preferentially occupied by Si 3p electrons. Where the states near to the conduction band could express the dominating characteristics of Mg 3s and 3p electrons. Also, it is observed that the S value is gradually increasing with increasing temperature till 550°C. On the other hand, comparing with the undoped Mg_2Si, the addition of Bi 0.010 at.% in Mg_2Si potentially reduces the S value. Increasing Bi concentration ($x = 0.015$, 0.020 and 0.025 at.%) further reduces the S value. The superior S value of 348 µV/K is recorded for intrinsic Mg_2Si and the lower value of 135 µV/K for Mg_2Si doped with Bi 0.025 at.% at absolute temperature 550°C. Remarkably, the increase in Bi doping concentration ($x = 0.010–0.025$ at.%) in Mg_2Si causes the increase in dislocation defects which could be understood from the interplanar shift and lattice parameter (Figure 5.10a1). Furthermore, the nano polycrystalline grains significantly enhance the grain boundaries and its

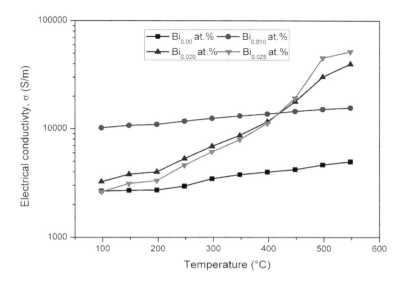

FIGURE 5.15 Electrical conductivity of $Mg_2Si_{1-x}Bi_x$ compound [28].

intergrain junctions. All these phenomena are well reflected in the strong variation in the crystallographic texture of different planes resulting in the anisotropic nature in parental Mg_2Si. The evolved polycrystalline texture results in scattering at grain boundaries resulting from the reduction in the charge carrier mobility in addition to an ionized impurity scattering due to Bi doping. This occurrence can be represented by an empirical relation $S \approx r - \ln nc$, where r denotes the scattering factor and nc represents the carrier concentration which significantly influenced the reduction in the S value.

Figure 5.15 shows the temperature dependence electrical conductivity behavior of Mg_2Si doped with Bi (0.010–0.025 at.%). The undoped Mg_2Si exhibits an intrinsic semiconducting behavior. This is due to the band gap of 0.7 eV. Where in, Mg_2Si doped with Bi (0–0.025 at.%) is significantly contributed by the electrons (affirmed by the negative polarity in Figure 5.14) as the higher carrier concentration. An increase in the temperature favorably increases the electrical conductivity of Mg_2Si samples with the increase in Bi concentration ($x = 0.010$–0.025 at.%) as compared with undoped Mg_2Si. The higher electrical conductivity of 52×10^3 S/m is recorded for Mg_2Si doped with $Bi_{0.025}$ at.% and the undoped Mg_2Si of 5.71×10^3 S/m. This is due to higher crystal imperfections due to higher Bi (0.025.%) doping and physical defects including dislocation and porosities. This is well understood from the corresponding strong and higher evolution of anisotropic texture in bulk Mg_2Si doped with $Bi_{0.025}$ at.%. LoBotz et al. reported the temperature-dependent carrier mobility in $Mg_2Si_{1-x}Bi_x$ by $\mu \alpha T - 3/2$ and acoustic lattice scattering phenomena. It is mathematically expressed as by Fiameni S. et al. [34]:

$$\sigma = en\mu \frac{e^2 n \tau}{m^*} \tag{5.8}$$

Therefore, the carrier concentration due to intrinsic semiconducting characteristics remarkably influences by reducing carrier mobility by electron-photon scattering. In spite, the increment in electrical conductivity with the increase in temperature could be seen for the Bi-doped condition, the σ plots pertaining to 0.020 and 0.025 at.% doped Mg_2Si exhibit the gradual increase in semiconduction up to 400°C.

After that, higher σ phenomena are noticed. From the earlier reports, it is concluded by both computational and experimental approaches, the doping of Bi more than 0.020 at.% is observed to be excess doping due to its solubility limits in Mg_2Si reported earlier [35]. It is predominantly affected by the narrowing and overlapping of conduction and valence bands resulting in the higher electrical conductivity phenomena.

Figure 5.16 shows the total thermal conductivity behavior of Mg_2Si doped with Bi. It is understood that the plots confirm that Bi doping of increasing concentration from 0.005 to 0.025 at.% in Mg_2Si significantly reduces the K value. The undoped Mg_2Si possesses 5.74 W/m-K at 100°C that is linearly reduced to 4.38 W/m-K at 550°C. However, the peak reduction of K value is recorded with 2.69 W/m-K at 550°C that is corresponding to B 0.020 at.% doped Mg_2Si.

This has led to achieving the enhancement in the ZT of Mg_2Si Bi doped with Bi than the intrinsic Mg_2Si (Figure 5.17). The higher ZT of 0.66 at 550°C in Mg_2Si doped with Bi 0.020 at.% is higher than the undoped Mg_2Si having the ZT of 0.11. The remarkable reduction is concurrently attributed to the factors such as an increase in the functional grain boundaries and interfaces by nanostructure features and defects (by doping and porosities).

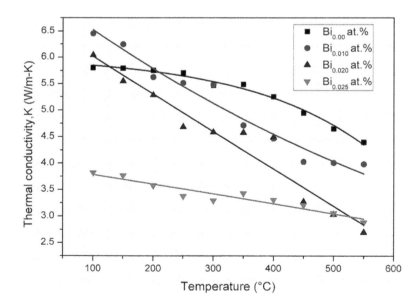

FIGURE 5.16 Thermal conductivity characteristics of $Mg_2Si_{1-x}Bi_x$ compound.

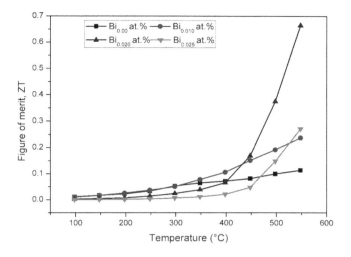

FIGURE 5.17 Figure of merit characteristics of $Mg_2Si_{1-x}Bi_x$ compound.

5.6 CONCLUSIONS

The synthesis and simultaneous densification of Mg_2Si doped with Bi thermoelectric compound have been done by mechanical milling and spark plasma sintering. The following significant observations have arrived.

Mechanical milling of Mg-Si with Bi results in nanostructuring and intimate mixing of elemental powders (Mg, Si, and Bi). The varying Bi doping concentration plays a significant role in crystallite size reduction and corresponding lattice strain in the Mg-Si mixture during milling. The nanostructuring and intimate mixing results in enhancing reactive enthalpy, which aids superior reactivity during SPS. The combustible Mg-Si with Bi doping undergoes reaction synthesis and simultaneous densification of superior relative density more than 96%. The crystallite in bulk Mg_2Si shows the average crystallite size in the range of 43–47 nm with a negligible variation. EBSD grain orientation map shows that the bulk Mg_2Si possess the uniform and finer grain size with anisotropy texture evolution in the negligible range with the increase in Bi doping concentration. The undoped Mg_2Si shows good enhancement in the σ to higher-order 103. The σ for Mg_2Si doped with $Bi_{0.025}$ at.% shows the σ of 52×10^3 S/m. Notably, the nano-crystallite features, point defects due to interstitial doping, and the porosities have significantly helped to lower the total thermal conductivity close 2 W/m-K at 550°C from the 5.65 W/m-K observed in the intrinsic Mg_2Si to achieve a higher ZT of 0.66 for Mg_2Si doped with $Bi_{0.020}$ at.%.

REFERENCES

1. Energy ststistics report, Govt. of India, 2014.
2. A.F. Ioffe, L.S. Stilbans, E.K. Iordanishvili, T.S. Stavitskaya and A. Gelbtuch, Semiconductor Thermoelements and Thermoelectric Cooling. *Physics Today* 12, 5 (1959), 42. doi:10.1063/1.3060810.

3. F.W. Dynys, A. Sayir, J. Mackey, A. Sehirlioglu, Thermoelectric properties of WSi_2-Si_xGe_{1-x} composites, *J. Alloys Compd.* 604 (2014), 196–203.
4. I-J. Chen, A. Burke, A. Svilans, H. Linke, C. Thelander, Thermoelectric power factor limit of a 1D nanowire, *Phys. Rev. Lett.* 120 (2018), 177703.
5. R. Murugasami, P. Vivekanandhan, S. Kumaran, R. Suresh Kumar, T. John Tharagan, Synergetic enhancement in thermoelectric and mechanical properties of n-type SiGe alloys prepared by solid state synthesis and spark plasma sintering, *Mater. Res. Bulletin* 118 (2019), 110483.
6. IEA global energy and CO_2 status report, 2017.
7. D.K. Aswal, R. Basu, A. Singh, Key issues in development of thermoelectric power generators: High figure-of-merit materials and their highly conducting interfaces with metallic interconnects. *Energy Convers. Manag.* 114 (2016), 50–67.
8. G.S. Nolas, J. Sharp, J. Goldsmid, *Thermoelectrics: Basic Principles and New Materials, 173, Developments*, Springer, 2001. doi: 10.1007/978-3-662-04569-5.
9. P. Vivekanandhan, R. Murugasami, K.V.R.S. Sairam, S. Kumaran Densification and mechanical properties of nanostructured Mg_2Si thermoelectric material by spark plasma sintering, *Pow. Technol.* 319 (2017), 129–138.
10. R.T. Littleton, T.M. Tritt, J.W. Kolis IV, Semiconductors and semimetals. In: *Recent Trends in Thermoelectric Materials Research*, Vol. 70, edited by T.M. Tritt, Academic Press: New York, 2000, pp. 77–115.
11. K. Nielsch, J. Bachmann, J. Kimling, H. Böttner, Thermoelectric nanostructures: From physical model systems towards nanograined composites, *Adv. Energy Mater.* 1 (2011), 713–731.
12. A. Willfahart, Screen printed thermoelectric devices, Linköpings Universitet, SE60174 Norrköping, 2014.
13. A.M. Dehkordi, M. Zebarjadi, J. He, T.M. Tritt, Thermoelectric power factor: Enhancement mechanisms and strategies for higher performance thermoelectric materials, *Mater. Sci. Eng. R* 97 (2015), 1–22.
14. A. Shakouri, Recent developments in semiconductor thermoelectric physics and materials, *Annu. Rev. Mater. Res.* 41 (2011), 399–431.
15. D.M. Rowe, C.M. Bhandari, *Modern Thermoelectric*, Reston Publishing Company, 1983.
16. G.A. Slack, in: M. Rowe (Ed.), *CRC Handbook of Thermoelectrics*, CRC Press: Boca Raton, FL, 1995.
17. D.M. Rowe, *CRC Handbook on Thermoelectric*, CRC Press, 1995.
18. V.E. Borisenko, A.B. Filonov, *Semiconducting Silicides*, Springer-Verlag: Berlin Heidelberg, 2000.
19. V.E. Borisenko, *Semiconducting Silicides: Basics, Formation, Properties*, Springer, Series in Materials Science, 1st ed., 2000.
20. P. Vivekanandhan, R. Murugasami, S. Appu Kumar, S. Kumaran, Structural features and thermoelectric properties of spark plasma assisted combustion synthesised magnesium silicide doped with Al, *Mater. Chem. Phy.* 241 (2020), 122407.
21. E. Godlewska, K. Mars, R. Mania, S. Zimowski, Combustion synthesis of Mg_2Si, *Intermetallics* 19 (2011), 1983–1988.
22. F. Predel (2016) Thermodynamic properties of Mg-Si (magnesium-silicon) system. In: *Phase Equilibria, Crystallographic and Thermodynamic Data of Binary Alloys*. Physical Chemistry, Vol 12D, edited by F. Predel, Springer: Berlin, Heidelberg.
23. C. Yang, M.D. Zhu, X. Luo, L.H. Liu, W.W. Zhang, Y. Log, Z.Y. Xiao, Z.Q. Fu, L.C. Zhang, E.J. Lavernia, Influence of powder properties on densification mechanism during spark plasma sintering, *Scripta Mater.* 139 (2017), 96–99.
24. R. Marder, C. Estourne`s, G. Chevallier, R. Chaima, Plasma in spark plasma sintering of ceramic particle compacts, *Scripta Mater.* 82 (2014), 57–60.

25. J-i. Tani, H. Kido, Thermoelectric properties of Bi-doped Mg$_2$Si semiconductors, *Physica B* 364 (2005), 218–224.
26. Z. Trzaska, G. Bonnefont, G. Fantozzi, J-P. Monchoux, Comparison of densification kinetics of a TiAl powder by spark plasma sintering and hot pressing, *Acta Mater.* 135 (2017), 1–13.
27. Z. Trzaska, A. Couret, J-P. Monchoux, Spark plasma sintering mechanisms at the necks between TiAl powder Particles, *Acta Mater.* 118 (2016), 100–108.
28. P. Vivekanandhan, R. Murugasami, S. Kumaran, Rapid in-situ synthesis of nanocrystalline magnesium silicide thermo-electric compound by spark plasma sintering, *Mater. Lett.* 197 (2017), 106–110.
29. P. Vivekanandhan, R. Murugasami, S. Kumaran, Microstructure and mechanical properties of nanocrystalline magnesium silicide thermoelectric compound prepared via Spark plasma assisted combustion synthesis, *Mater. Lett.* 231 (2018), 109–113.
30. R.T. Li, Z.L. Dong, K.A. Khor, Spark plasma sintering of Al–Cr–Fe quasicrystals: Electric field effects and densification mechanism, *Scripta Mater.* 114 (2016), 88–92.
31. V. Tong, J. Jiang, A.J. Wilkinson, T. BenBritton (2015), The effect of pattern overlap on the accuracy of high resolution electron backscatter diffraction measurements, *Ultramicroscopy* 155, 62–73.
32. S-W. You, H. Kim, Solid-state synthesis and thermoelectric properties of Bi-doped Mg$_2$Si compounds, *Curr Appl Phys* 11 (2011), S392–S395.
33. P. Nierodaa, J. Leszczynskia, A. Kolezynski, Bismuth doped Mg$_2$Si with improved chomogeneity: Synthesis, characterization and optimization of thermoelectric properties, *J. Phy. Chem Solids* 103 (2017), 147–159.
34. S. Fiameni, S. Battiston, S. Boldrini, A. Famengo, F. Agresti, S. Barison, M. Fabrizio, Synthesis and characterization of Bi-doped Mg$_2$Si thermoelectric materials, *J Solid State Chem* 193 (2012), 142–146.
35. J-i. Tani, H. Kido, Thermoelectric properties of Bi-doped Mg$_2$Si, semiconductors, *Physica B*, 364 (2005), 218–224.

6 The Role of Computational Intelligence in Materials Science: An Overview

J. Ajayan and Shubham Tayal
SR University

Sandip Bhattacharya
Hiroshima University

L. M. I Leo Joseph
SR University

CONTENTS

6.1 Introduction 125
6.2 Prediction of Critical Temperature in Superconductors 127
6.3 Flaw Detection in 3D Composite Materials 128
6.4 Pharmaceutical Tableting Processes... 128
6.5 Processing of Smart Laser Materials.. 129
6.6 Determination of Energy Bandgap of Semiconductors 131
6.7 Lattice Constant Prediction of Perovskites.. 132
6.8 Adsorption Isotherms of Metal Ions onto Zeolites................................... 133
6.9 Material Constants of Composite Materials 134
6.10 Conclusion 136
References... 136

6.1 INTRODUCTION

Computational intelligence represents the capability of a machine or a computer or a system to understand a special task from the available information or experimental data [1–3]. The main objective of computational intelligence techniques is to create a model for analyzing and processing data of human beings. Computational intelligence techniques use different computational algorithms to obtain hidden information from available data and use this information or knowledge for training a machine or a computer to take complicated decisions without the help of human beings. CNN (convolutional neural network) and ANN (artificial neural network)-based computational models can be used for predicting the mechanical and physical properties of a material [4]. In 2019, Alireza Nasiri et al. [5] reported that the acoustic emission events and the deep learning-based CNN model can be used for the online monitoring of damage of SiC_f–SiC_m composite materials. This online monitoring of damage of SiC_f–SiC_m composite materials is important for ensuring the safety of nuclear reactor systems. Computational intelligence models can also be used for the optimization of polymer electrolyte fuel cells (PEFCs), which are considered one of the most attractive renewable energy sources for meeting future global energy needs [6,7].

Real-time recognition of materials is a fundamental issue in the field of computational intelligence [8] due to the variations in illuminations, light conditions, and camera perspectives. Machine learning can play a key role in solid-state materials science especially for predicting the stability of materials, structure of materials, and estimation of material properties [9]. There are mainly three types of machine learning techniques, namely unsupervised learning, supervised learning, and reinforcement learning. Among these, supervised learning is widely used in materials science and its workflow is depicted in Figure 6.1. The success of supervised learning

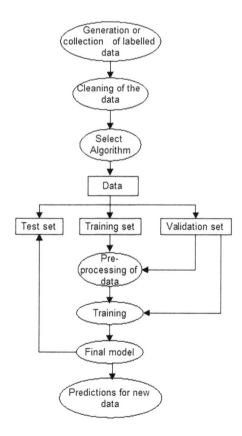

FIGURE 6.1 Workflow of supervised learning [9].

depends on the quality and quantity of data available [10]. The rapid advancement of computational intelligence techniques brings a new dawn into the growth of materials science. Data collection, data representation, and selection of computational algorithms are the key elements of computational intelligence techniques [11,12].

6.2 PREDICTION OF CRITICAL TEMPERATURE IN SUPERCONDUCTORS

The prediction of critical temperatures (T_C) of superconducting materials is extremely important because they are widely used in applications such as power systems, controllable nuclear fusion, transportation, cutting-edge scientific equipment, medical instruments, etc. Machine learning and deep learning techniques have been widely used for the prediction of material properties in the last several years. These prediction models use elementary properties of materials as original data input. In 2020, Yabo Dan et al. [13] reported that computational intelligence can play a key role in predicting the T_C of materials having superconductivity characteristics. The convolutional gradient (CG) boosting decision trees (BDT) model can be used to predict the T_C of superconductors. CNNs are found to be suitable for learning the hierarchical features from the original data sets. CNN-based computation models have been widely used to enhance the characterization method after modeling the microstructure data of materials to predict the molecular and crystal structure properties. CNN-based prediction models depend on their hierarchical features that are usually associated with fully interlinked layers for classification or regression. For small data sets, these interlinked layers should not be too complicated or too simple. However, the CG-BDT model is robust compared with CNN-based models.

A database is essential to test and train the prediction models. Therefore, the SuperCon data set [14] can be used to train the CG-BDT model prediction of T_C of superconductors. Learning algorithms and data representations are the heart of any machine learning-based prediction models. Crystal structure and molecular formula are the two major material representation methods. For molecular formula-based material representation, chemical composition can be taken as an input whereas crystal structure-based material representation requires the development of vector-based crystal structures. One-hot coding and Magpie descriptors are the two most popularly used material representation methods. The CG-BDT algorithm can be computed as a boosting technique depending upon the decision trees as

$$f_m(x) = f_{m-1}(x) + \sum_{j=1}^{J} C_{mj} \quad (6.1)$$

$f_m(x) = m$th learner
$C_{mj} =$ loss of the jth node on the mth tree

Forward distribution algorithms can be used to train the CG-BDT prediction model. The process of the CG-BDT prediction model is illustrated in Figure 6.2. The CG-BDT prediction model has two parts, namely, a convolutional feature extraction and regression prediction. Data pre-processing and model training are essential for this type of prediction algorithms.

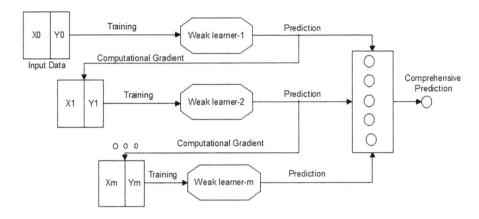

FIGURE 6.2 The process of the CG-BDT prediction model [13].

6.3 FLAW DETECTION IN 3D COMPOSITE MATERIALS

In 2019, Yongmin Guo et al. [15] reported the detection of flaws in 3D braided composite materials using a fully CNN/GRU (Gated Recurrent Unit) model. Anti-ablation capability, through-thickness reinforcement, and high damage tolerance are the major advantages of 3D braided composite materials over traditional laminated composite materials. Therefore, these materials are widely used in a large number of industries. The detection of defects in these materials is important in improving the reliability and life of these materials. The fully CNNs consist of six temporal convolutions, which usually act as feature extractions and the GRU block consists of two layers that improve the classification of fully CNNs. Ultrasonic signals are given as input data sets in these models.

6.4 PHARMACEUTICAL TABLETING PROCESSES

In the pharmaceutical industry, it is important to utilize the knowledge of the relationship between the critical material attributes (CMA) and product quality for the development of new products and manufacturing process optimization. Computational intelligence models can be used to identify the critical process parameters (CPP), CMA, and critical quality attributes (CQA). Particle swarm optimization (PSO), genetic algorithm (GA), social spider optimization algorithm (SSO), flower pollination algorithm (FPA), Cuckoo search algorithm (CSA), BAT algorithm, and Grey Wolf optimization (GWO) algorithms are the widely used seven bio-inspired optimization algorithms in developing computational intelligence models for pharmaceutical tableting processes. In 2018, H. M. Zawbaa et al. [16] developed a bio-inspired optimization algorithms-based computational intelligence model for pharmaceutical tableting processes. ANN can be used as a regression model for prediction. The block diagram of the computational intelligence model used in pharmaceutical tableting processes is shown in Figure 6.3. Average feature extraction, mean square error, and standard deviation are the three key parameters that can be used to evaluate the prediction performance of the computational intelligence model shown in Figure 6.3.

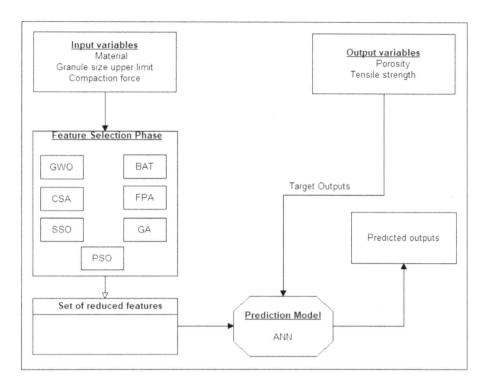

FIGURE 6.3 The block diagram of the computational intelligence model used in pharmaceutical tableting processes [16].

6.5 PROCESSING OF SMART LASER MATERIALS

Computational intelligence involves the usage of different computational algorithms to capture the hidden information from the available input data and use this captured information to train the intelligent machine to make complex decisions without the help of human beings. Computational intelligence can play a key role in the processing of laser materials for high-power laser sources such as excimer (126–351 nm), neodymium-doped YAG lasers (1064 nm), and semiconductor lasers (800–1000 nm). Over the years, the laser power keeps on increasing while the power consumption is decreasing. The processing steps of all laser materials can be precisely controlled by a proper combination of interaction time and power density. The processing of laser materials can be mainly classified into three types and they are [17]

  I. Heating w/o melting or vaporizing like in surface hardening
 II. Melting w/o vaporizing like in cladding
III. Vaporizing like in cutting

Computational intelligence involves the usage of computing techniques for characterization, modeling, and forecasting in materials science engineering [18]. The widely used computational intelligence techniques are fuzzy logic (FL), genetic algorithm (GA), and ANNs. Among these, ANNs are the most widely used computational intelligence technique. ANNs are self-learning systems. ANNs serve like a biological brain whose fundamental unit is a neuron. The model of a neuron unit is shown in Figure 6.4.

$x_1, x_2...x_N$ = Signals
$w_1, w_2...w_N$ = weights

The representation of a multilayer ANN is depicted in Figure 6.5.

An ANN consists of an input layer, an output layer, and one or more hidden layers. The weights are usually computed during the training phase. During the training phase, process parameters will be used as forward stimulus. A back propagation error signal will be generated by taking the difference between the actual result and the network output. The weights will be regularly updated based on these back propagation error signals. The process will be repeated until the error is zero or minimum. Batch back propagation, incremental back propagation, and Levenberg–Marquardt back propagation are the widely used back propagation learning methods. Back propagation is a supervised learning technique that needs a known desirable output for every input data during the learning phase. There are two types of neural networks namely static neural networks and dynamic neural networks. The architecture of a fuzzy logic system is shown in Figure 6.6.

The key components of fuzzy logic are

1. Fuzzifier
2. The intelligence engine
3. Defuzzifier
4. Rules (knowledge base module)

The fuzzifier converts the crisp system inputs into fuzzy sets. IF-THEN rules are stored in the RULES block. The intelligence module performs the reasoning process with the help of rules. Finally, the defuzzifier converts the fuzzy sets back to crisp numbers. Takagi–Sugeno–Kang and Mandeny fuzzy rules are the two most popularly used fuzzy rules. The fuzzy logic-based laser material processing flow chart is given in Figure 6.7. The GA and ANN-based laser material processing flow charts are given in Ref. [19].

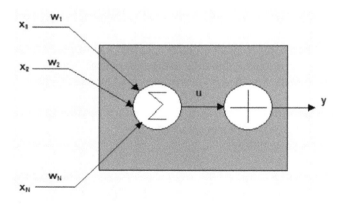

FIGURE 6.4 The model of a neuron unit [19].

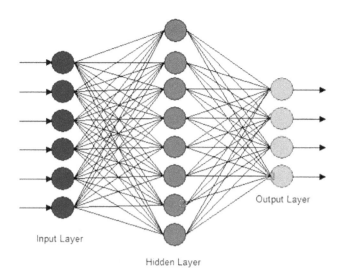

FIGURE 6.5 Representation of a multilayer ANN [19].

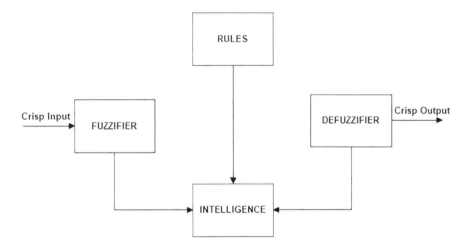

FIGURE 6.6 The architecture of a fuzzy logic system [19].

6.6 DETERMINATION OF ENERGY BANDGAP OF SEMICONDUCTORS

In 2016, T. O. Owolabi et al. [20] reported the calculation of band gap energy (E_g) of ZnO semiconductors using computational intelligence techniques. ZnO is a II-VI direct band gap compound semiconductor. An SVR (support vector regression) computational intelligence model was used to determine the E_g of a doped ZnO semiconductor. The SVR computational intelligence technique is developed by Vladimir. Structural risk minimization inductive principles are used in SVR algorithms to obtain good generalization on

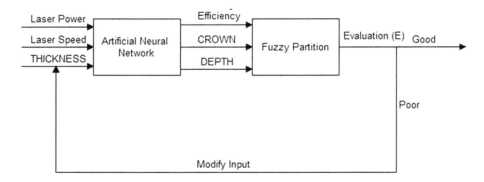

FIGURE 6.7 The fuzzy logic-based laser material processing flow chart [19].

limited learning patterns. It utilizes kernels and does not use local minima. In order to develop an SVR algorithm for the determination of the band gap of doped ZnO semiconductors, the lattice constants (*a* and *c*) of the semiconductor must be mapped with the help of Gaussian kernel function onto *k*-dimensional feature space (Eq. 6.2):

$$f(E_g, \omega) = \sum_{i=1}^{k} \alpha_i g_i (x = a \,\&\, c) + b \qquad (6.2)$$

$g_i(a, c)$ = set of nonlinear transformations
b = bias term

A loss function can be used to evaluate the quality of determination of E_g of the ZnO material. In this case, the loss function can be described by

$$L(E_g, f(E_g, \omega)) = 0 \text{ if } |E_g - f(E_g, \omega)| \leq \varepsilon$$

$|E_g - f(E_g, \omega)| - \omega$.

Otherwise, RMSE (root mean square error) can be used to evaluate the accuracy of the SVR-based band gap prediction model. GA- and ANN-based computational intelligence techniques can be used for the growth of quantum dots and their band gap prediction [21].

6.7 LATTICE CONSTANT PREDICTION OF PEROVSKITES

Lattice constant plays a key role in the identification of materials and also to estimate their material properties. Perovskites are widely used materials in the development of solar cells. Quantum mechanics-based density functional theory computational tools can be used to predict the physical and chemical material properties. Multiple linear regression-based computational models have also been developed to predict the properties of perovskites. SVR, random forest, and GRNN (generalized regression neural network) models can be used to predict the lattice constants of perovskites. The flow chart for the development of the computational intelligence prediction model is shown in Figure 6.8.

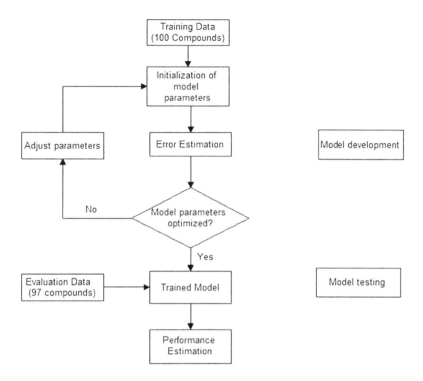

FIGURE 6.8 The flow chart for the development of the computational intelligence prediction model [22].

6.8 ADSORPTION ISOTHERMS OF METAL IONS ONTO ZEOLITES

A large amount of industrial effluents are released into soils and water resources, leading to harmful effects on the environment and human health. The industrial effluents contain a large quantity of toxic metallic ions like Ag^+, Co^{2+}, and Cu^{2+} that need to be removed before releasing into soil and water. Cu^{2+} ions can cause severe nervous system issues, liver damage, kidney damage, and mucosal irritations in human beings. The Cu^{2+} ions also affect the growth of fish in water, invertebrates, and aquatic plants. Silver contamination in water leads to severe health problems such as diarrhea, vomiting, and abdominal pain. Co^{2+} contamination in water can lead to bone defects, diarrhea, and lung irritation. Techniques such as solvent extraction [23], electrochemical methods [24], ion exchange resins [25,26], and chemical precipitation can be used for the removal of the above-mentioned toxic metallic ions. But these techniques are very costly. In this scenario, the adsorption technique has emerged as an attractive solution for the treatment of industrial effluents that contain toxic metal ions due to their ease of operation and low cost. In 2020, M. S. Netto et al. [27] reported the use of ANN and ANFIS (adaptive neuro-fuzzy interface systems)-based computational intelligence models for predicting the adsorption isotherms of metal ions such as Ag^+, Co^{2+}, and Cu^{2+} onto different zeolites (ZSM-5, ZH+, and Z4A). The special structural features of zeolites make them a suitable metallic ion adsorbent. The concentration of solutions, molecular weights of metal ions, Si/Al ratios of zeolites, and temperatures can be taken as input variables of ANN and ANFIS-based computational intelligence models for the prediction of adsorption isotherms. The equilibrium adsorption capacity of Ag^+, Co^{2+}, and Cu^{2+} ions can be taken as target variables.

6.9 MATERIAL CONSTANTS OF COMPOSITE MATERIALS

In 2014, W. Beluch et al. [28] reported a two-scale method to predict the composite material constants. Gradient methods and ANNs can be used for determining the material constants of composite materials. The block diagram of a distributed evolutionary algorithm (DEA) that can be used for the identification of material constants of composite materials is depicted in Figure 6.9. In the DEA, the entire individual

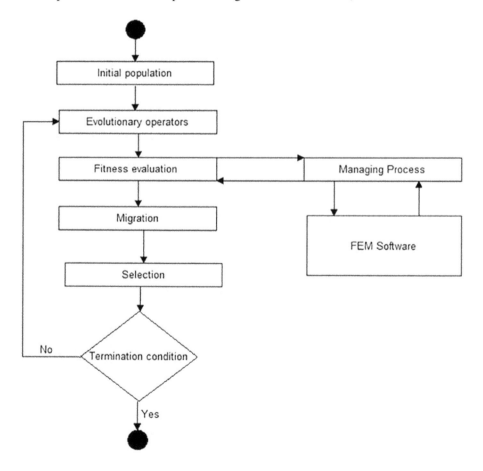

FIGURE 6.9 The block diagram of distributed evolutionary algorithm (DEA) [28].

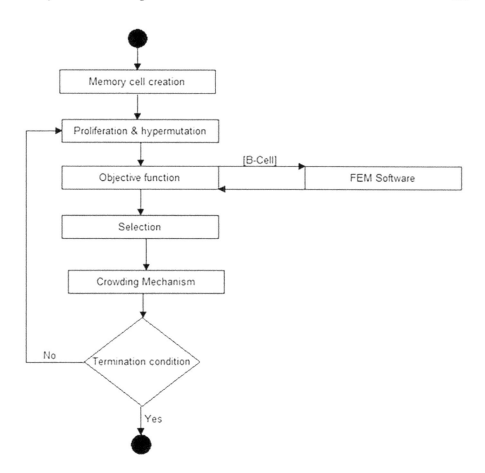

FIGURE 6.10 The block diagram of parallel artificial immune system (PAIS) [28].

population will be divided into two or more sub-populations with the same number of individuals in each sub-population. Each sub-population works independently and exchanges information with one another during the migration phase. The block diagram of PAIS is shown in Figure 6.10.

The artificial immune systems (AIS)-based algorithms work like the biological immune system of mammals. The PAIS operation shuts with the randomly created set of memory cells that undergo proliferation. The clone numbers are based on the objective function values. The clones hypermutate to generate B-cells. B-cells replace several memory cells during the selection process that takes place on the basis of the distance between B-cells and memory cells. To eliminate the identical memory cells, a crowding mechanism can be used. The two-scale approach (refer [28]) for the material constant identification of composite materials is depicted in Figure 6.11.

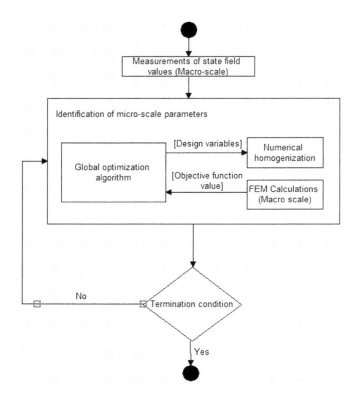

FIGURE 6.11 Two-scale approach for the material constant identification of composite materials [28].

6.10 CONCLUSION

Computational intelligence is gaining enormous attention over the years in the field of materials science. This chapter covered an overview of the role of computational intelligence in the field of materials science. The computational intelligence models can be used for determining the critical temperatures of superconductors, detection of flaws in 3D-composite materials, pharmaceutical tableting processes, processing of smart laser materials, predicting the energy band gap and lattice constants of semiconductors, and also predicting the adsorption performance of metallic ions into zeolites and so on.

REFERENCES

1. R.S. Peres, X. Jia, J. Lee, K. Sun, A.W. Colombo, J. Barata, Industrial artificial intelligence in industry 4.0- systematic review, challenges and outlook, *IEEE Access*, Vol. 8, pp. 220121–220139, 2020.
2. Y. Pan, L. Zhang, Roles of artificial intelligence in construction engineering and management: A critical review and future trends, *Automation in Construction*, Vol. 122, p. 103517, 2021.

3. A.F.S. Borges, F.J.B. Laurindo, M.M. Spínola, R.F. Gonçalves, C.A. Mattos, The strategic use of artificial intelligence in the digital era: Systematic literature review and future research directions, *International Journal of Information Management*, Vol. 57, p. 102225, 2021.
4. D. Merayo, A. Rodríguez-Prieto, A.M. Camacho, Prediction of physical and mechanical properties for metallic materials selection using big data and artificial neural networks, *IEEE Access*, Vol. 8, pp. 13444–13456, 2020.
5. A. Nasiri, J. Bao, D. Mccleeary, S.M. Louis, X. Huang, J. Hu, Online damage monitoring of SiCf-SiCm composite materials using acoustic emission and deep learning, *IEEE Access*, Vol. 7, pp. 140534–140541, 2019.
6. J. Ajayan, D. Nirmal, P. Mohankumar, M. Saravanan, M. Jagadesh, L. Arivazhagan, A review of photovoltaic performance of organic/inorganic solar cells for future renewable and sustainable energy technologies, *Superlattices and Microstructures*, Vol. 143, p. 106549, 2020.
7. Y. Chen, B. Peng, Multi-objective optimization on multi-layer configuration of cathode electrode for polymer electrolyte fuel cells via computational-intelligence-aided design and engineering framework, *Applied Soft Computing*, Vol. 43, pp. 357–371, 2016.
8. H. Zhang et al., Gathering effective information for real-time material recognition, *IEEE Access*, Vol. 8, pp. 159511–159529, 2020.
9. J. Schmidt et al., Recent advances and applications of machine learning in solid-state materials science, *npj Comput Mater*, Vol. 5, p. 83, 2019.
10. D. Morgan, R. Jacobs, Opportunities and challenges for machine learning in materials science, *Annual Review of Materials Research*, Vol. 50, Issue 1, pp. 71–103, 2020.
11. W. Sha, Y. Guo, Q. Yuan, S. Tang, X. Zhang, S. Lu, X. Guo, Y. Cao, S. Cheng, Artificial intelligence to power the future of materials science and engineering, *Advanced Intelligent Systems*, Vol. 2, p. 1900143, 2020.
12. L. Himanen, A. Geurts, A.S. Foster, P. Rinke, Data-driven materials science: Status, challenges, and perspectives, *Advanced Science*, Vol. 6, p. 1900808, 2019.
13. Y. Dan, R. Dong, Z. Cao, X. Li, C. Niu, S. Li, J. Hu, Computational prediction of critical temperatures of superconductors based on convolutional gradient boosting decision trees, *IEEE Access*, Vol. 8, pp. 57868–57878, 2020.
14. Z. Yu, F. Liu, R. Liao, Y. Wang, H. Feng, X. Zhu, Improvement of face recognition algorithm based on neural network, in *Proceedings of 10th International Conf. Measuring Technol. Mechtron. Automat. (ICMTMA)*, Feb. 2018, pp. 229–234.
15. Y. Guo et al., Fully convolutional neural network with GRU for 3D braided composite material flaw detection, *IEEE Access*, Vol. 7, pp. 151180–151188, 2019.
16. H.M. Zawbaa, S. Schiano, L. Perez-Gandarillas, C. Grosan, A. Michrafy, C-Y. Wu, Computational intelligence modelling of pharmaceutical tableting processes using bio-inspired optimization algorithms, *Advanced Powder Technology*, Vol. 29, Issue 12, pp. 2966–2977, 2018.
17. J. Dutta Majumdar, I. Manna, Laser material processing, *International Materials Reviews*, Vol. 56, Issue 5–6, pp. 341–388, 2011.
18. L.A. Dobrzanski, J. Trzaska, A.D. Dobrzanska-Danikiewicz, Use of neural networks and artificial intelligence tools for modeling, characterization, and forecasting in material engineering, *Comprehensive Materials Processing*, Vol. 2, pp. 161–198, 2014.
19. G. Casalino, Computational intelligence for smart laser materials processing, *Optics & Laser Technology*, Vol. 100, pp. 165–175, 2018.
20. T. Owolabi, M. Faiz, S.O. Olatunji, I.K. Popoola, Computational intelligence method of determining the energy band gap of doped ZnO semiconductor, *Materials & Design*, Vol. 101, pp. 277–284, 2016.

21. A.P. Singulani, O.P. Vilela Neto, M.C. Aurélio Pacheco, M.B.R. Vellasco, M.P. Pires, P.L. Souza, Computational intelligence applied to the growth of quantum dots, *Journal of Crystal Growth*, Vol. 310, Issue 23, pp. 5063–5065, 2008.
22. A. Majid, A. Khan, T-S. Choi, Predicting lattice constant of complex cubic perovskites using computational intelligence, *Computational Materials Science*, Vol. 50, Issue 6, pp. 1879–1888, 2011.
23. S. Padungthon, M. German, S. Wiriyathamcharoen, A.K. Sengupta, Polymeric anion exchanger supported hydrated Zr(IV) oxide nanoparticles: A reusable hybrid sorbent for selective trace arsenic removal, *Reactive & Functional Polymers*, Vol. 93, pp. 84–94, 2015.
24. I. Heidmann, W. Calmano, Removal of Zn(II), Cu(II), Ni(II), Ag(I) and Cr(VI) present in aqueous solutions by aluminium electrocoagulation, *Journal of Hazardous Materials*, Vol. 152, pp. 934–941, 2008.
25. B. Alyüz, S. Veli, Kinetics and equilibrium studies for the removal of nickel and zinc from aqueous solutions by ion exchange resins, *Journal of Hazardous Materials*, Vol. 167, pp. 482–488, 2009.
26. S. Rengaraj, K.H. Yeon, S.H. Moon, Removal of chromium from water and wastewater by ion exchange resins, *Journal of Hazardous Materials*, Vol. 87, pp. 273–287, 2001.
27. M.S. Netto, J.S. Oliveira, N.P.G. Salau, G.L. Dotto, Analysis of adsorption isotherms of Ag^+, Co^{+2}, and Cu^{+2} onto zeolites using computational intelligence models, *Journal of Environmental Chemical Engineering*, Vol. 9, Issue 1, p. 104960, 2021.
28. W. Beluch, T. Burczyński, Two-scale identification of composites' material constants by means of computational intelligence methods, *Archives of Civil and Mechanical Engineering*, Vol. 14, Issue 4, pp. 636–646, 2014.

7 Characterization Techniques for Composites using AI and Machine Learning Techniques

Dr. Zeeshan Ahmad
Al Falah University

Dr. Hasan Zakir Jafri
Bharat Institute of Engineering and Technology

Dr. R. Hafeez Basha
Bhasha Research Corporation

CONTENTS

7.1	Introduction	140
7.2	Microscopy	140
	7.2.1 Optical Microscopy	141
	7.2.2 Scanning Electron Microscopy (SEM)	143
	7.2.3 Transmission Electron Microscopy (TEM)	146
	7.2.4 Scanning Tunneling Microscopy (STM)	148
	7.2.5 Field Ion Microscopy	149
	7.2.6 Atomic Force Microscopy (AFM)	151
7.3	Spectroscopy	153
	7.3.1 Ultraviolet Spectroscopy (UV-Spectroscopy)	153
	7.3.2 Fourier Transform Infrared Radiation Spectroscopy (FTIR-Spectroscopy)	154
	7.3.3 Optical Emission Spectroscopy (OES)	156
	7.3.4 Energy Dispersive Spectroscopy (EDS)	159
	7.3.5 X-ray Diffraction (XRD)	159

7.4	AI and ML Techniques for Material Characterization	161
	7.4.1 Feature Mapping Using the Scale-Invariant Feature Transform (SIFT) Algorithm	161
	7.4.2 Image Classification Using Machine Learning and Deep Learning	163
7.5	Conclusion	163
References		164

7.1 INTRODUCTION

Characterization is the key requirement in the development of new composite materials to understand their structure and behavior. It is a basic process in the field of composite materials to understand the relationship of composition to structure and probed properties. The scientific understanding of composite materials could be ascertained without characterization. The characterization of composite materials is done on two scales; one is microscopic characterization in which microscopic structural analyses are carried out by using different techniques such as scanning electron microscopy, X-ray diffraction, etc. and the second is macroscopic characterization in which mechanical, physical, and tribological behaviors are studied. This chapter emphasizes on microstructure characterization of new developing composites.

Numerous characteristic techniques such as optical microscopy, scanning electron microscopy, atomic force microscopy, and Fourier transform infrared spectroscopy have been exercised to characterize the composites. These techniques are used to characterize the structure of composites including bulk structure, surface structure, surface topology, and microstructure and to characterize the composition of composites consisting of bulk composition, surface composition, and impurities as well as the local environment (Salame et al., 2018). These characterization techniques are also comprised in two categories as microscopy and spectroscopy based on application. Microscopy is a technique that is used to explore and map the surface and subsurface of a composite. It is based on employing photons, electrons, and ions to collect the information of a sample surface. Spectroscopy techniques are used to expose the chemical composition and its variation, crystal structure, and photoelectric properties of composites. These techniques are based on various scientific principles. Various characterization techniques are shown in Figure 7.1.

7.2 MICROSCOPY

Microscopy is a technique that is used to view an object or area of object at a high resolution by using lenses. In this technique, an electromagnetic radiation or electron beam is used to interact with the specimen to get the information of material that is under investigation. An electromagnetic radiation or electron beam incident on the specimen of a material and the signal-produced specimen are collected to create an image. There are several microscopic techniques available for the characterization of composite materials. In this chapter, we discuss few microscopic techniques that are generally used for the characterization of composites.

Characterization Techniques for Composites 141

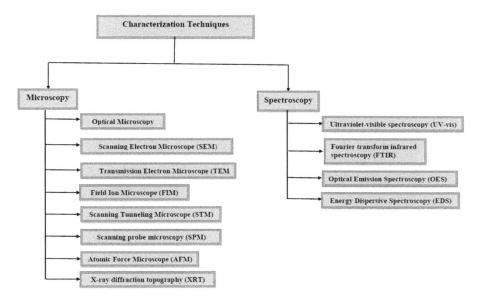

FIGURE 7.1 Various characterization techniques.

7.2.1 Optical Microscopy

Optical microscopy is a technique used to magnify images of a sample and characterize the microstructure by using one or a series of lenses with visible light. This has glass or plastic lenses, which are curved that reflect rays of light and can magnify objects. The illumination of an optical microscope is based on the reflection of visible light. The basic principle of an optical microscope is shown in Figure 7.2. The visible light passes through the condenser lens, which is focused on a half polished mirror and diverted toward the objective lens. The function of an objective lens is to focus these rays on the specimen. After falling on the specimen, it will go back again toward the objective lens and produce an inverted intermediate image of the specimen. These rays again go from an intermediate image to eyepiece lens which produces the magnified image of the specimen.

Magnification of the image can be achieved by increasing the number of lenses. But the resolution of the image deteriorates as the magnification of the image increased. The resolution of an optical microscope is defined as the minimum spacing between two points which can noticeably be observed through the microscope to be separate entities or features. The resolution or resolving power of an optical microscope is given by Rayleigh's criteria. According to Rayleigh's criteria, the resolution limit r_1 is given as

$$r_1 = \frac{0.61\,\lambda}{\mu \sin \alpha}$$

So, the resolution of the optical microscope depends on the wavelength of light, half aperture angle, and refractive index of the lens material. Depth of focus is also an

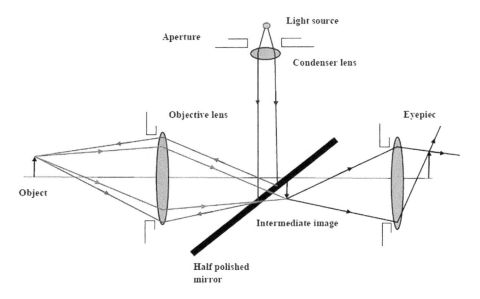

FIGURE 7.2 Principle of an optical microscope.

important parameter related to an optical microscope. Depth of field is an important parameter associated with the objective lens of an optical microscope and is defined as the limit of position for the object for which our eye can notice without any variation in the sharpness of the image. The depth of field h is given as

$$h = \frac{1.22\,\lambda}{\mu \sin\alpha \tan\alpha}$$

The depth of field is also dependent on the wavelength of the light, refractive index of the lens material, and half aperture angle. So, the depth of field of an optical microscope can be varied by adjusting the aperture angle and the resolution of the image decreased as the depth of field increased.

Field of focus is also an important parameter associated with the eyepiece and is defined as the limit of positions at which the image can be viewed without appearing out of focus, for a fixed position of the object. The change in the distance of image formation dv is given as

$$dv = -M^2 h$$

where M is the magnification and h is the field of depth. Field of focus in an optical microscope states a plane on which a clear image of the specimen is formed.

The image created by an optical microscope depends on the contrast of sample features such as brightness, phase, color, polarization, and fluorescence based on the illumination source. So little amount of surface preparation such as polishing and etching is required to make one side reflected and get the best suitable results (Mukhopadhyay, 2003). Putri et al. (2019) identified the phase present

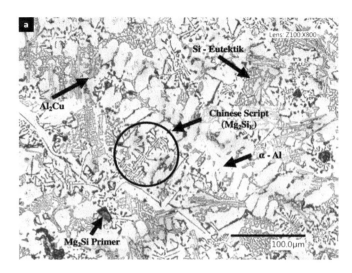

FIGURE 7.3 Phase identification in a composite microstructure using an optical microscope with 500× magnification (Putri et al., 2019).

in the development of a composite and observed the porous structure formed due to various chemical reactions with the help of an optical microscope as shown in Figure 7.3. Optical microscopy is a technique used to define the characteristics of new composite materials such as composite constituent cracks and volume fraction, particle size, and voids content (Yen et al., 1991). Sharp et al. (2019) used an optical microscope to determine the fiber orientation of an elliptical fiber. Joyce et al. (1997) developed a technique to characterize the localized fiber waviness in the case of laminate composites.

7.2.2 Scanning Electron Microscopy (SEM)

An optical microscope has a limitation of maximum magnification up to 1000X. This magnification power is not only limited by the number of lenses or refractive property of the lens material but also restricted by the wavelength of light used. An electron microscope was developed due to the restricted wavelength of light. In an electron microscope, a beam of electrons is used because electrons have a much shorter wavelength, which enables us to get better resolution. The wavelength of the electron beam is based on potential difference and is given by the De Broglie equation as:

$$\lambda = \frac{h}{\sqrt{2\, m_o eV}}$$

where λ is the wavelength, h is the Planck constant, e is the charge, and m_o is the rest mass of an electron. A scanning electron microscope is a type of electron microscope used to characterize and investigate the surface through a focused beam of electrons and produce high-resolution images of the sample. SEM can give information on

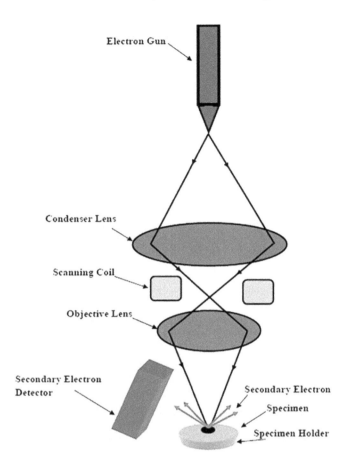

FIGURE 7.4 Schematic view of scanning electron microscopy.

the surface topology, crystalline structure, and chemical composition of a composite sample. The basic working principle of a scanning electron microscope is shown in Figure 7.4. A focused beam of electrons generated through an electron source at the top of the column accelerates under vacuum; in order to prevent the strike of an electron with any atoms and molecules present in the column, electrons of the focused beam are controlled by the use of electromagnetic lenses. Condenser lenses are used to focus the electron beam onto the sample while the main function of the objective lens is to define the resolution of images. Scanning coils are employed to raster the beam on the sample. This electron beam interacts with the atoms present in the specimen and various types of electrons such as the secondary electron, backscattered electron, and characteristic X-rays are emitted. These signals are collected with the help of a two-electron detector. The image that appears on the screen represents contrast information between areas with various chemical compositions.

The electron beam falls on the surface of the specimen and penetrates the surface of the specimen to a certain depth depending on the accelerating voltage and density of the specimen. The electrons are accelerated due to the potential difference V; the kinetic energy of accelerated electron equals potential energy and is mathematically defined as follows:

$$v = \sqrt{\frac{2\ eV}{m_o}}$$

where e is the charge, m_o is the rest mass of the electron, and v is the velocity of electron, and so the acceleration of electron is also dependent on potential difference only. The maximum resolving power of a scanning electron microscope depends on many factors such as the size of a spot by electron beam and the volume of electrons interacting with the specimen. In the case of SEM, special surface preparation is required to get the best results of the specimen. Surface preparation is done in order to enhance the electrical conductivity of the specimen, so it can withstand a high-energy electron beam. A non-conductive specimen accumulates the charge when an electron beam strikes the specimen causing image artifacts. So the specimen for scanning electron microcopy must be conductive, at least the surface. Non-conductive materials are generally coated with an ultrathin coating of an electrically conductive material such as gold, platinum, tungsten, and iridium in order to achieve a good result. The limitation of a scanning electron microscope is that it cannot give atomic resolution.

The surface and internal morphology of a specimen up to 1000Å or better resolution can get through the scanning electron microscope with special surface preparation (Mukhopadhyay, 2003). SEM techniques find many applications in the characterization of various types of composites. Ahmad et al. (2020) used a scanning electron microscope to analyze the microstructure of silicon nitride reinforced LM 25 composites to find the dispersion and clustering of reinforcement. The microstructure of aluminum alloy LM 25 shows solution of Al and inter-dendritic network of aluminum silicon eutectic as shown in Figure 7.5. The mapping of fractured surfaces in the tensile test was also carried out through SEM in order to identify the fracture mechanism. SEM is a technique capable of giving a three-dimensional (3D) image with great resolution and is used to characterize the morphology of the sample surface, particle size, microorganism, and fragments (Rydz et al., 2019). Antony et al. (2015) carried out scanning electron microscopic analysis of aluminum-based hybrid

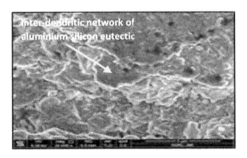

FIGURE 7.5 SEM micrograph of aluminium alloy LM 25 (Ahmad et al., 2020).

composites at different magnifications to evaluate the clustering and porosity present in developed composites. SEM analysis of aluminum-based composite microstructures shows fractional composition of reinforcement based on the chemical composition of a matrix material and grain structure of reinforcement (Nishchev et al., 2013). Porras and Maranon (2012) used the SEM technique to determine the fracture mechanism in laminated composites. SEM analysis can provide a rapid and versatile analysis of conductive polymer nanocomposites with various percentages of reinforcement (Dikin et al., 2006).

7.2.3 Transmission Electron Microscopy (TEM)

TEM is also a type of electron microscopy. The working of TEM is the same as SEM, but in the case of TEM, the electrons that strike on the specimen are transmitted and collected by a detector after passing through a projective lens as shown in Figure 7.6. These striking electrons are interacted with molecules of the specimen and transmitted through it and give valuable information about the internal microstructure. The specimen prepared for TEM is very ultrathin so electrons can easily transmit through it. The resolving power of TEM is 2–5 Å, which is far greater than SEM as 1000 Å (Mukhopadhyay, 2003). So, TEM gives much better information about the internal microstructure of the specimen while SEM gives information about the surface and its composition only. The valuable information such as crystal structure, morphology, and stress state about the new developing composite and present material can easily get through TEM. So, TEM is capable to provide high-resolution 2D images and give valuable information on microstructure.

The main problem with TEM is the surface preparation of the specimen. The complex step of surface preparation is to make an ultrathin specimen because cutting a specimen in ultrathin form can change the structure of the specimen. So in this case, the information gathered through TEM is artifacts. Generally, the specimen of TEM is prepared by an electropolishing process. In the case of a metal matrix composite, the electropolishing process is able to make quickly a thin foil so the electron can easily transmit from the metal matrix. Cairney et al. (2000) proposed an effective process as the ion beam miller process to prepare the specimen for a metal matrix composite. Qing et al. (2015) also proposed three preparation processes such as focused ion beam, ion beam etching, and ultramicrotomy for carbon fiber reinforced composites and suggested that these methods are useful to get an in-depth understanding at the interphase. This technique is also used to identify flaws, fractures, and damage also. Preparing the specimen is highly laborious, and the chances to change the structure of the specimen during preparation are the drawbacks of this technique.

Appiah et al. (2000) used the TEM technique to investigate interfacial metal matrix laminated composites reinforced with carbon fiber. The microstructure of the composite was observed in the TEM image (Figure 7.7) in which the layers of matrix clearly consist of carbon and silicon carbide from the fiber. This technique provides a very distinct feature of carbon–carbon interfacial that enables a very accurate analysis of the microstructure of carbon fiber and differentiates the carbon coating on fiber. This technique is also used to evaluate the particle size of

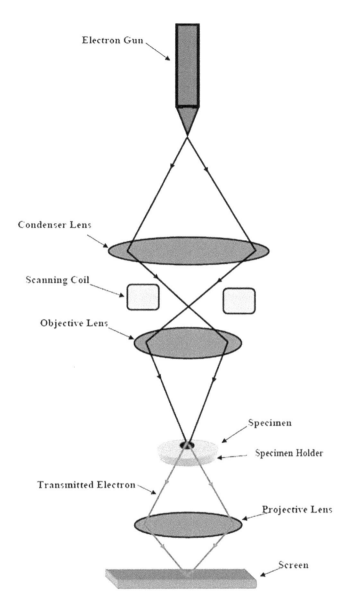

FIGURE 7.6 Schematic view of transmission electron microscopy.

formed nanodispersed particles of organic compounds in polymer matrix composites (Pastukhov & Davankov, 2011). The dislocation arrangement in the metal matrix can be easily observed subjected to fatigue loading through TEM (Hancock, 1967). Montealegre-Melendez et al. (2017) identified precipitant morphology and evaluated the shape reinforcement of titanium-based composites reinforced with boron carbide by using TEM.

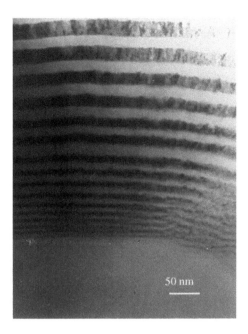

FIGURE 7.7 TEM image of a cross-section of a composite (Appiah et al., 2000).

7.2.4 SCANNING TUNNELING MICROSCOPY (STM)

STM is a non-optical electron-based technique, which is used to evaluate the atomic level microstructure of a metallic surface. This technique is able to provide a three-dimensional image and characterize the topology of the specimen. This technique is based on quantum tunneling in which a very sharp metal tip is brought very close in the range of few angstroms to a specimen (Figure 7.8) and a bias voltage is applied between the specimen and the tip allowing the electrons to tunnel through the vacuum gap between them, resulting in the flow of a tunneling current. This tunneling current $I(x)$ is calculated as follows:

$$I(x) = \text{constant} \times eVe^{\left[-2\frac{\sqrt{2m\delta}}{h}d\right]}$$

where e is the electron, m is the mass of the electron,; V is the bias voltage,; δ is the height of the barrier, h is Planck's constant, and d is the distance between the tip and the specimen. The tunneling current is dependent on the local density of the state of the specimen and applied bias voltage, and it is exponentially dependent on the distance between the tip and the specimen. A very sharp tip like a needle has been treated as a single atom projected from its end mounted on a piezoelectric tube which experiences a mechanical deformation. The tunneling current is kept constant during the operating condition. The tip is scanned over the surface of the specimen and its position is changed by a piezoelectric actuator in order to keep the tunneling current constant. The image is produced which contains information about the geometry

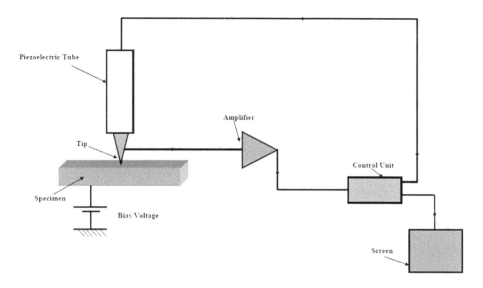

FIGURE 7.8 Schematic view of scanning tunneling microscopy.

of the structure, surface roughness, and defects, and the size and conformation of molecules, and their aggregation at the surface of the specimen are evaluated. The resolution of the image is restricted by the curvature radius of the tip. The chance of image artifact is possible if the tip is made other than that single atom at the end. The challenges of this technique are to make an extremely clean surface of the specimen, a very sharp tip, and vibration control. The tip is generally made of platinum–iridium or tungsten through mechanical sharing and thermochemical etching.

Novoselov et al. (2015) analyzed the 2D atomic crystal structure of various materials through STM. Caballero-Quintana et al. (2020) evaluated the morphology, energetic level alignment of thin solid films of low molecular weight molecules after applying thermal annealing at liquid-solid phase through STM. Nakamura (2018) used the STM technique to characterize the silicon nanodots produced by the ultrathin silicon oxide film technique as shown in Figure 7.9. This technique is limited to conducting specimens.

7.2.5 Field Ion Microscopy

Field ion microscopy is a technique used to characterize metal and semiconductor specimens at atomic resolution. It is used to observe dislocation in atomic structure and grain boundaries and also provide information about the particle size, shape, volume fraction, etc. The basic principle of field ion microscopy is shown in the figure. In this technique, a sharp tip is produced and placed on an electrically insulated stage and in a highly ultra vacuum chamber, which is filled with a gas such as helium or neon (Figure 7.10). This tip is cooled at a cryogenic temperature such as 20–100 K and a positive voltage of 5–10 KV is applied on it. Due to the applied charge on the tip, the surrounding area ionized. The ions absorbed by the tip are also ionized and

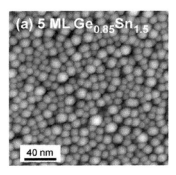

FIGURE 7.9 STM micrograph of five-monolayer (ML) Ge0.85Sn0.15 nanodots (Nakamura, 2018).

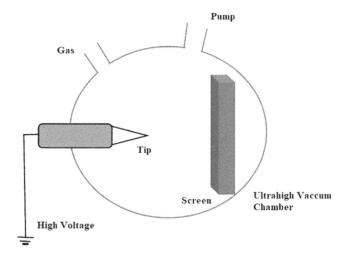

FIGURE 7.10 Schematic view of field ion microscopy.

turn positively charged, thus repelling from it. These repelled ions are absorbed by a detector and produce an image of good resolution of atomic structure.

This technique can be provided the resolution of the range of 1 Å which is much higher than light-based microscopy. The characterization of materials through field ion microscopy is limited by the materials which can be produced in tip form and tolerate high electrostatic filed as well as high vacuum condition. So this technique is restricted to refractory metals such as tungsten, tantalum, iridium, and rhenium, which can survive a strong electric field at the tip without evaporation from the surface. The tip is generally prepared by an electrochemical polishing process in which the chances to form asperities also enhance. The formations of asperities cause artifacts in the image of the specimen. The application of this technique is found to be very less in the field of composite materials due to a large number of restrictions. Sauvage et al. (2005) characterized copper–iron composites by field ion microscopy. FIM analyses (as shown in Figure 7.11) were performed to investigate the clustering

Characterization Techniques for Composites

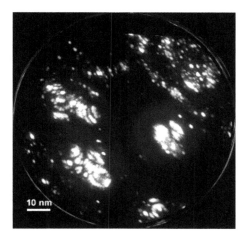

FIGURE 7.11 FIM micrograph of a copper–iron composite (Sauvage et al., 2005).

and determine the grain size of iron particles. The results indicate a little amount of clustering with the grain size of 5–50 nm.

7.2.6 ATOMIC FORCE MICROSCOPY (AFM)

AFM is a technique used for evaluating the material properties and imaging the surface at high resolution up to the atomic level. It has the ability to measure the local properties of materials such as friction, electrical force, resistance, capacitance, and conductivity. The advantage of this technique is that it can be imagined almost on any kind of surfaces like polymers, ceramics, glass, composites, etc. The basic working principle of AFM is shown in Figure 7.12. AFM uses a laser beam system to

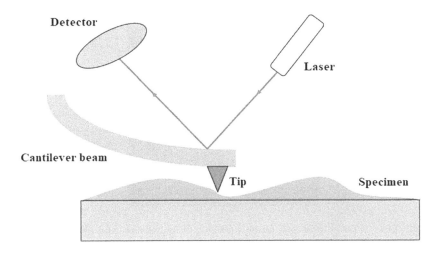

FIGURE 7.12 Basic principle of atomic force microscopy.

measure the properties and analyze the surface of the specimen. It uses a cantilever beam with a sharp tip to scan the surface of the specimen. The tip and cantilever beam are fabricated by microsilicon or silicon nitride and the tip is of pyramid shape and has dimensions of 3–6 µm and 15–40 nm end radius. The vertical and lateral deflection of the beam is used to evaluate the surface of the specimen through the laser beam reflected off the cantilever. The deflection of the cantilever beam is monitored by tracking this reflected laser through a detector.

AFM can be used to image various types of surfaces such as ceramics, polymers, glass, composites, and biological surfaces. It provides information of both kinds as quantitative and qualitative on various properties like size, morphology surface texture, roughness, and micro-/nano-coating. This can also be used to evaluate and localize different forces such as adhesion strength, magnetic forces, and mechanical properties. Downing et al. (2000) suggested that AFM is an influential technique to evaluate the interphase due to a lower interaction force between specimen and probe and is used to determine the size of interphase and stiffness of glass fiber reinforced epoxy polymer composites. Wang (1996) characterized interphase region properties of fiber-reinforced polymer composites through AFM. AFM is a powerful tool to evaluate surface morphology due to its ability to create three-dimensional images at angstrom and nanoscale and is used to determine the dispersion of nanometric fillers, morphology of nanocomposites, and polymer blend based elastomers. Sausa and Scuracchio (2014) and Durate et al. (2020) characterized multiwalled nanotubes in metal matrix composites through AFM. The mapping of stiffness of all developed composites was carried out through AFM as shown in Figure 7.13. Gawdzinska et al. (2014) and Senthilbabu et al. (2015) used AFM to evaluate the surface topology and surface roughness of metal matrix composites. Evaluation of average grain size and oxides phase tracing in the case of an oxide reinforced metal matrix composite can be easily done through AFM (Koleukh et al., 2018).

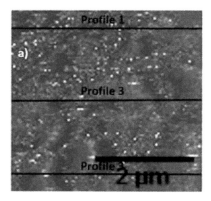

FIGURE 7.13 AFM micrograph at the interface of magnesium and MCWNTs (Durate et al., 2020).

7.3 SPECTROSCOPY

Spectroscopy is a technique based on the interaction of electromagnetic radiation with matter in different ways and this information derived is used for the characterization of small and large molecules of matter. It is a quantum phenomenon and it depends on both, properties of radiation and the structural composition of a material which is interacted with electromagnetic radiation. Electromagnetic radiation is defined as a form of energy whose origin is the emission and absorption of charged particles. The energy of electromagnetic radiation is expressed as follows:

$$E = \frac{hc}{\lambda}$$

$$E = h\nu$$

where h = Planck's constant; c = velocity of light; λ = wavelength of radiation; and ν = frequency of radiation in Hz.

In spectroscopy where electromagnetic radiation falls on the material specimen which is under investigation, molecules of the material get excited and move to a higher energy level after absorption of energy of this falling electromagnetic radiation. During the transition state, the molecules of material also emit light and the intensity of emitted light is less than the incident light. This emitted light is used to derive the information of the material to be characterized.

The spectroscopy is classified based on the type of radiative energy used, nature of the interaction, and type of material used for characterization. Type of energy distinguished by the wavelength region of the spectrum includes microwave, infrared, ultraviolet-visible, X-ray, etc. The nature of the interaction is classified based on the interaction between energy and material that occurs such as absorption, emission, elastic scattering, and inelastic scattering. Spectroscopy is also designed based on the interaction of radiant energy with a specific type of matter such as atoms, molecules, crystals, and nuclei. In this chapter, we discuss a few types of spectroscopy techniques that are generally used for the characterization of composite materials.

7.3.1 Ultraviolet Spectroscopy (UV-Spectroscopy)

UV-spectroscopy is absorption-type spectroscopy in which an electromagnetic energy wave of the UV region (200–400 nm) is used to interact with the specimen. Generally, a light source of tungsten filament lamp or hydrogen-deuterium lamp is used for an electromagnetic wave in spectroscopy (Figure 7.14). A monochromator composed of a prism and slit is used to obtain the UV-region. In the monochromator, the prism dispersed the emitted radiation and the UV-region is selected by using slit. This monochromatic beam is passed through the specimen of the material to be characterized. The molecules of the specimen containing π-electron or non-bonding electron absorbed the energy of falling monochromatic beam and get excited from the molecule ground level to a higher anti-bonding energy level. After absorption of electromagnetic energy by the specimen, it creates a different spectrum which will

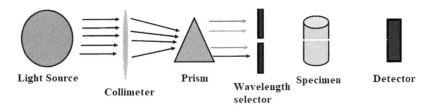

FIGURE 7.14 Schematic view of working of UV-spectroscopy.

give the information regarding identifying the elements and quantitive concentration of elements present in the specimen. The elemental concentration in a specimen is given by Beer–Lambert's law:

$$A_0 = \log_{10}\left(\frac{I_L}{I_T}\right) = \alpha c L$$

where A_0 = measured absorbance; I_L = intensity of incident light at a given wavelength; I_T = transmitted light; α = extinction co-efficient for each element and wavelength; c = concentration of the absorbing element; and L = path length through the specimen.

UV-spectroscopy is mainly used for polymer-based composites. This can be used for analyses of intermolecular interaction of composites and to find the optical properties of polymer-based composites. This technique is not generally preferred for biomaterials characterization because biomaterials are not in a solution and the extinction coefficient of various biomaterials is unknown. Ananda et al. (2016) identified the absorption characteristics of polymer polyvinyl alcohol-based composites with respect to sodium sulfate concentration through UV-spectroscopy. The optical characterization of ZnO dispersed particles in PANI can also be done by UV-spectroscopy (Kayathri et al., 2013). Lu et al. (2007) used UV-spectroscopy to determine the size of ZnO nanoparticles dispersed in polymer-based composites. The absorbance of bare ZnO particles in ethanol solution can also be done by UV-spectroscopy as shown in Figure 7.15. UV-spectroscopy is also used to determine the interfacial morphology, interparticle spacing, and finite size of nanoparticles and their effect on optical properties of developed composites (Srivastava et al., 2008). Yan et al. (2019) characterized silver nanoparticles reinforced polymer-organic based composites.

7.3.2 Fourier Transform Infrared Radiation Spectroscopy (FTIR-Spectroscopy)

FTIR-spectroscopy is also referred to as absorption-type spectroscopy in which infrared electromagnetic radiation interacts with the matter. Infrared electromagnetic radiation includes radiation of a large wavelength than visible light or it carries less energy as well as less wave number (cm^{-1}). It is further subdivided into near mid and far spectroscopy based on wavelength. Near-infrared spectroscopy has large energy as compared to mid and far-infrared spectroscopy and it can penetrate more

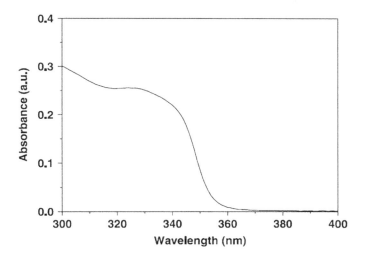

FIGURE 7.15 UV-absorption spectrum of ZnO particles in ethanol (Lu et al., 2007).

but it is less sensitive. In this technique, a number of wavelength infrared beams are created and incident on the specimen simultaneously and each wavelength is examined individually.

FTIR spectroscopy is based on the Michelson interferometer experimental as shown in Figure 7.16. The infrared beam is obtained from a light source through a diffraction grating system and passed through an interferometer. The interferometer consists of three parts, a beam splitter, which has the ability to transmit half radiation incident on it and reflect the other half, a fixed mirror, and a moving mirror. The half-transmitted portion of radiation moves to the fixed mirror and the other half-reflected portion moves to the moving mirror. The beam reflected back and combined causing interference and finally incident on the specimen. This energy is absorbed by the molecules of the specimen and starts to move in vibratory motion. The motion may be a type of symmetric and unsymmetric stretching motion, deformation or bending, other types of motion, etc. The portion of the incident beam transmitted through the specimen carries the actual information of the specimen which is under investigation and recorded with the help of a detector. The detector produces an electrical signal and this electrical signal formation acquired by an FTIR spectrometer is known as an interferogram. As the interferogram is collected, a Fourier transfer algorithm is applied to yield the reference spectrum which gives the quantitative information of the specimen. FTIR has significant advantages such as a better signal-to-noise ratio, faster acquisition of TR spectrometer, and superior wavelength accuracy. FTIR is performed to find out the types of bonding and chemical fingerprints as well as characterize new materials or identify and verify known and unknown specimens.

FTIR spectroscopy is generally used to identify organic, inorganic, and polymeric materials. Its application is also found to characterize the unknown materials, notice contaminates in a material, locate additive, and recognize decomposition and oxidation. Cecen et al. (2008) used FTIR spectroscopy to observe the interaction between matrix and reinforcement of polyester and epoxy-based composites. Kalaleh et al.

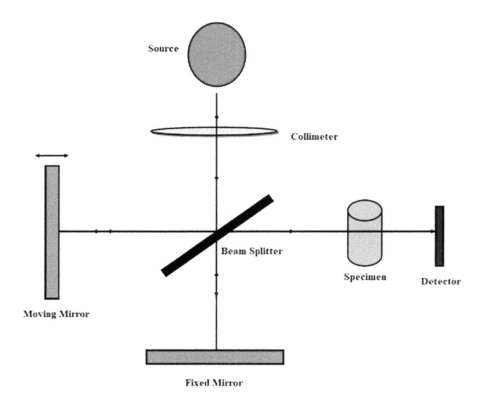

FIGURE 7.16 Schematic views of working of Fourier transform infrared radiation spectroscopy.

(2013) characterized polymer-based composites through FTIR spectroscopy. FTIR analyses of bentonite, a copolymer, and a composite material were carried out to understand the formation and crosslinking of hydrogel with the mixture. The FTIR spectrum of bentonite shown in Figure 7.17 indicates the peaks identified as OH deformation and stretching. Characterization of weathered wood and evaluation of the loss of wood particles after weathering can also be done by FTIR spectroscopy (Stark & Matuana, 2007). FTIR spectroscopy is used to analyze the thermal damage and contaminates of epoxy carbon composites (Rein & Seelenbinder, 2015; Seelenbinder, 2015). Weiss et al. (1997) used FTIR spectroscopy to characterize the organic or mineral composites used for bone and dentist substitute materials. Analysis of the interaction between polyamide 6 and organic material and structure of developed polymer-based composites was carried out by FTIR spectroscopy (Krylova & Dukstiene, 2019).

7.3.3 Optical Emission Spectroscopy (OES)

OES is an emission type spectroscopy in which an electromagnetic energy wave of the visible spectrum and part of UV-spectrum (130–800 nm) in terms of wavelength

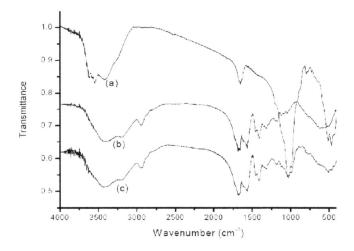

FIGURE 7.17 FTIR spectrum of bentonite (Kalaleh et al., 2013).

interact with the specimen. In this, an electrode with a high electrical source is used to heat the surface of the specimen up to some thousand degrees (Figure 7.18). An electrical discharge is produced due to the potential difference between the electrode and the specimen. This discharge passes through the specimen, heats up, and vaporizes as well as triggers the atoms of material as shown in Figure 7.18. These excited atoms emit the characteristic emission spectrum. The light created by discharge contains a collection of spectral lines. The light coming from vaporizing the surface of the specimen is known as plasma. This light is split into a number of specific elemental wavelengths through a diffraction grating system and detected by a detector. Each

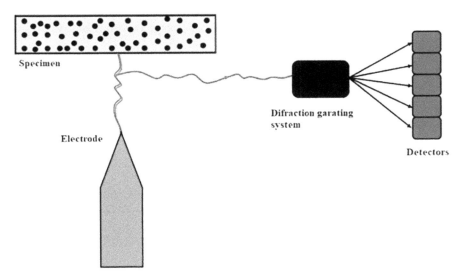

FIGURE 7.18 Basic working principle of optical emission spectroscopy.

atom of elements present in the specimen emits a unique spectrum and the intensity of the spectrum is proportional to the concentration of the element.

The emission spectrum is created because the electrons at the outer shell are ejected as they get electromagnetic energy and make the atom unstable. The electrons at the outer shell are easily ejected because of weak bonding due to the large distance from the nucleus. The electron comes from a higher shell to fill the vacancy and restore the stability of the atom. So, the transition of an electron from one energy level to another energy level occurs and releases an excess amount of energy. This excess energy creates a spectrum line. A series of spectrum lines are formed by the transition of different electrons of various elements to different energy levels. Each transition creates a unique optical emission line with a fixed wavelength or radiation energy.

OES is used to determine chemical composition and characterize the coating and filler materials in composites. Bousqet et al. (2009) used OES to characterize silicon nitride coating on a silicon carbide surface through argon–nitrogen plasma. Hanada et al. (2010) characterized the diamond or hydrogenized coating on composites through a plasma gun by OES. An optical emission spectrum of plasma generated by coaxial arc plasma deposition (CAPD) and pulsed laser deposition (PLD) is shown in Figure 7.19. Optical emission spectra were carried out at wavelengths from 300 to 850 nm for both cases. The spectrum was obtained in the form of peaks with various intensities which are formed from a different kind of C2 swan bond, C-atom,

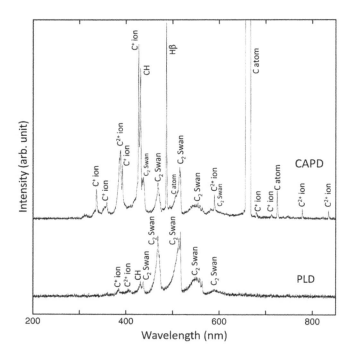

FIGURE 7.19 Optical emission spectra of plasma generated by CAPD and PLD (Hanada et al., 2010).

Characterization Techniques for Composites 159

and C-ions. Dijken et al. (1989) analyzed the chemical composition of filler materials of an inorganic filler in materials through OES. Veklich et al. (2019) analyzed the thermal radial distribution and electron density of arc plasma in copper–chromium electrode composites by OES.

7.3.4 ENERGY DISPERSIVE SPECTROSCOPY (EDS)

EDS is also known as energy-dispersive X-ray spectroscopy (EDX) and energy-dispersive X-ray analysis spectroscopy (EDXA). In this technique, the specimen of material is bombarded with electrons through a source such as a tungsten tip. These electrons interact with the sample and ionize the electrons of the inner shell, typically K, L, or M shell. As the electrons are ionized and ejected, the electrons from the outer shell drop down to fill the vacancy and release the characteristic X-rays. The energy associated with the X-ray is equal to the difference in the energy level of the transition electron. These emitted X-rays are collected by a detector which converts them into an electronic signal. A pulse processor is used to measure these electronic signals of the detector and evaluate the energy of the X-ray. The results of EDS come in the form of a graph, where the Y-axis represents counts and the X-axis represents the energy level of those counts. EDS software is used to identify the elements present in the sample by the analysis of the energy level of X-rays.

EDS is a qualitative and quantitative analytical technique used for the elemental analysis or chemical characterization of the specimen. Qualitative analysis is done by the identification of lines in the spectrum and quantitative analysis involves measurement of line intensities for each element present in the sample. It is a non-destructive type and can be performed on a variety of samples in various states like liquid, solid, powder, and wafer (Asaka et al., 2004). Typically, SEM instrumentation is integrated with an EDS system to provide a chemical analysis of features being observed in SEM displays. Ahmad et al. (2020) used SEM integrated with EDS for microstructural analysis of silicon nitride reinforced aluminum-based composites. The EDS mapping was done for base material, reinforcement material, and all developed materials with various percentages of reinforcement. The various elements and their chemical composition present in aluminum alloy LM 25 are shown in Figure 7.20. In this micrograph, various peaks are observed according to the elements present in the aluminum alloy LM 25; these peaks were identified, and quantitative analysis was carried out by using EDS software. Ba et al. (2020) analyzed the chemical composition of a phosphate–nickel–titanium composite coating on steel by EDS. The presence of phosphate and tungsten elements in a thin film of composite coating on steel was done by EDS (Hosseini et al., 2008). Lee et al. (2017) carried out microstructural analyses of polymer–biomacromolecules composites in solid state through EDS.

7.3.5 X-RAY DIFFRACTION (XRD)

XRD is a technique in which X-ray is used to interact with the specimen of the material to be characterized. X-ray is a high-energy wave with a short wavelength of 2–3 Å. In this technique, monochromatic X-rays are produced by a source and collimated to concentrate and pass through the specimen. The specimen is rotated with

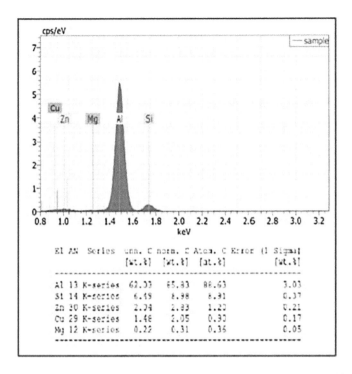

FIGURE 7.20 Energy dispersive micrograph of LM 25 alloy (Ahmad et al., 2020).

the help of a specimen holder through some angle and diffracted X-rays by the specimen are recorded through a detector. As the X-rays incident on the sample fulfill Bragg's equation, a constructive interference formed and a peak of intensity occur. Bragg's equation can be expressed as follows:

$$n\lambda = 2d \sin\theta$$

where n = positive integer; λ = wavelength of the incident wave; d = interplanar spacing of the crystal; and θ = angle of incident.

The diffraction peak is converted into a d-spacing and it will help to identify the element and phase of the element present in the specimen of the material as each element has a unique set of d-spacing. The resulting XRD analysis comes in the form of peak intensities versus angle and the peak intensity was identified by the use of JCPDS or other software. This is done by comparing d-spacing with standard reference patterns. XRD is a non-destructive technique designed to give more in-depth information about the identification and quantification of crystalline phases. Dobrzański et al. (2014) characterized an aluminum alloy A6061 composite reinforced with halloysite particles by using the XRD technique. The XRD technique is also used to analyze the morphology and crystallinity of polymer-based composites (Palanivel et al., 2017). Orbulov et al. (2008) used XRD to analyze the

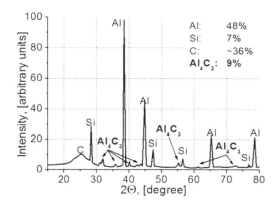

FIGURE 7.21 XRD micrograph of an aluminum matrix reinforced with zoltek-type carbon reinforcement (Orbulov et al., 2008).

interface between reinforcement and matrix of metal–matrix composites. In this research, various phases formed in composites and their distinguished planes were identified as shown in Figure 7.21. Wadatkar and Waghuley (2012) characterized polymer-based thin-film composite coatings through investigating the structure of polyvinyl acetate and determining the possible crystallinity in the composite film. Montealegre-Melende et al. (2018) identified the phase transformation and reaction that occurred during the sintering process of the Ti-TiAl-B_4C composites system by XRD. Patel et al. (2018) determined the crystallinity index, crystallite size, and average crystallinity of polymer-based composites. Strankowski et al. (2016) characterized polyurethane-based nanocomposites incorporated with reduced grapheme oxide by XRD.

7.4 AI AND ML TECHNIQUES FOR MATERIAL CHARACTERIZATION

7.4.1 Feature Mapping Using the Scale-Invariant Feature Transform (SIFT) Algorithm

This technique of feature mapping introduced by D. Lowe in 2004 from the University of British Columbia is invariant to image scale and rotation. Using this feature, some advantages can be drawn out for image processing used in AI and ML. The advantages are as follows:

Locality: this means that the features are local, so there is no need for occlusion and clutter.
Distinctiveness: each feature may be used to match from any database.
Quantity: for even a very small object, several features can be generated.
Efficiency: very high efficiency close to real-time performance.
Extensibility: can easily be extended to a wide range.

The algorithm
The SIFT algorithm has mainly five steps.
Scale-space peak selection
Keypoint localization
Orientation assignment
Keypoint descriptor
Keypoint matching

Scale-space
In the real world, the meaningfulness of the objects can be achieved at a certain scale. A glass full of water on a table might look perfect but has no meaning in terms of area equal to whole land on earth; in simple words, it does not exist. This type of nature related to objects is quite common. A scale-space attempts the same in digital images.

A scale-space is a function produced from the convolution of a Gaussian kernel at different scales in the input image. It is put into octaves which in turn depends on the size of the parent image. Each octave's size is half of the previous one.

Blurring
In each octave, images are blurred using Gaussian blur, which has an operator applied to each pixel. After this blurred images are used to generate another set of images. Difference of Gaussians (DoG) is then used for finding the interesting parts of an image.

Keypoint Localization
Key points, which are generated by Taylor series expansion, are used to get a more accurate location of extrema. A DoG has a better response as far as edges are concerned, so edges are not removed.

Orientation Assignment
For orientation assignment, we selected a neighborhood around the key point, to calculate gradient magnitude as well as the direction in that region. A histogram covering 360° is generated. The highest magnitude in the histogram having peak above 80% can be used for calculating orientation.

Keypoint descriptor
For the keypoint descriptor, a 16×16 window is divided into 16 sub-blocks of 4×4 size around the key point.

Keypoint Matching
Key points are matched using their nearest neighbors. It can use the nearest first match also due to noise.

7.4.2 IMAGE CLASSIFICATION USING MACHINE LEARNING AND DEEP LEARNING

Image classification is defined as the task of classifying "an image" from a set of given categories. There are several algorithms for machine learning such as:

K-Nearest Neighbour algorithm (KNN)
This method is used for classifying the objects on closest training in a given feature space. The training in this algorithm consists only of storing features and labels of the images. The classification takes place by assigning labels on testing examples. These algorithms are simple and can deal with multiclasses. The biggest disadvantage is using entire features for similarity computing leading to errors.

Support Vector Machine (SVM) algorithm
SVM is putting the examples as points in space; they are mapped by separating (dividing) a plane for maximizing the margin in between. The advantage of SVM is its capability for non-linear classification.

Ensemble Learning algorithm
Ensemble learning is used to aggregate a group and utilize them for the prediction where the most votes are cast.

Multi-layer Perceptron (MLP)
In MLP, there can be different or multiple layers between the input and the output including non-linear layers, known as hidden layers. MLP has great capability for non-linear models. However, it has a non-convex loss function and tuning for hidden layers.

Convolutional neural networks (CNNs)
CNNs have established themselves as master algorithms in the field of computer vision (Chollet, 2017). They merge the basic ideas of the architectures like Google Net, etc. but have inception modules with a special layer called a depth-wise separable convolution (Liu et al., 2018). In traditional convolutions, learning takes place simultaneously while in the separable convolution layer split feature and learning take place in two steps. In the first step, a single spatial filter is applied, while the second one takes care of cross-channel patterns.

7.5 CONCLUSION

In this chapter, various types of characterization for composite materials have been discussed for different types of composites. The authors have focused only on the microstructure of newly developed composites. The applications, advantages, and restrictions of characterization techniques with some research studies have been discussed. The detailed study on microstructure characterization will help future researchers to analyze the materials in detail and to choose the state-of-the-art techniques used in image processing and machine learning for automation of characterization based on various inputs obtained through characterization techniques.

REFERENCES

Ahmad, Z., Khan, S., Hasan, S. 2020. Microstructural characterization and evaluation of mechanical properties of silicon nitride reinforced LM 25 composite. *Journal of Materials Research and Technology* 9: 9129–9135. https://doi.org/10.1016/j.jmrt.2020.06.037

Ananda, H.T., Urs, G.T., Somashekar, R. 2016. Preparation and characterization of conductive PVA/Gly: Na_2SO_4 polymer composites. *Polymer Bulletin* 73: 1151–1165. doi: 10.1007/s00289-015-1540-z

Antony, M.D., Prakash, A., Jaswin, M.A. 2015. Microstructural analysis of aluminium hybrid metal matrix composites developed using stir casting process. *International Journal of Advances in Engineering* 1: 333–339.

Appiah, K.A., Wang, Z.L., Lackey, W.J. 2000. Characterization of interfaces in C fiber-reinforced laminated C–SiC matrix composites. *Carbon* 38: 831–838.

Asaka, Y., Miyazaki, M., Aboshi, H., Yoshida, T., Takamizawa, T., Kurokawa, H., Rikuta, A. 2004. EDX fluroscence analysis and SEM observations of resin composites. *Journal of Oral Science* 46: 143–148.

Ba, K., Chahine, A., Touhami, M.E., Alauzun, J.G., Manseri, A. 2020. Preparation and characterization of phosphate-nickel-titanium composite coatings obtained by sol–gel process for corrosion protection. *SN Applied Science* 2: 350. https://doi.org/10.1007/s42452-020-2173-x

Bousquet, A., Spinelle, L., Cellier, J., Tomasella, E. 2009. Optical emission spectroscopy analysis of Ar/N_2 plasma in reactive magnetron sputtering. *Plasma Processes and Polymers* 6: S605–S609. doi: 10.1002/ppap.200931602

Caballero-Quintana, I., Romero-Borja, D., Maldonado, J., Nicasio-Collazo, J., Amargós-Reyes, O., Jiménez-González, A. 2020. Interfacial energetic level mapping and nano-ordering of small molecule/fullerene organic solar cells by scanning tunneling microscopy and spectroscopy. *Nanomaterials* 10: 1–14. doi: 10.3390/nano10030427

Cairney, J.M., Smith, R.D., Munroe, P.R. 2000. Transmission electron microscope specimen preparation of metal matrix composites using the focused ion beam miller. *Microscopy and Microanalysis* 6: 452–462. doi: 10.1007/s100050010048

Cecen, V., Seki, Y., Sarikanat, M., Tavman, I.H. 2008. FTIR and SEM analysis of polyester- and epoxy-based composites manufactured by vartm process. *Journal of Applied Polymer Science* 108: 2163–2170.

Chollet, 2017. https://arxiv.org/pdf/1610.02357.pdf.

Dijken, J.W.V., Wing K.R., Ruyter, I.E. 1989. An evaluation of the radiopacity of composite restorative materials used in Class I and Class I1 cavities. *Acta Odontologica Scandinavica* 47: 401–407.

Dikin, D.A., Kohlhaas, K.M., Dommett, G.H.B., Stankovich S., Ruoff, R.S. 2006 Scanning electron microscopy methods for analysis of polymer nanocomposites. *Microscopy and Microanalysis* 12: 674–675. doi: 10.1017/S1431927606067080

Dobrzański, A.L., Tomiczek, B., Pawlyta M., Nuckowski, P. 2014. TEM and XRD study of nanostructured composite materials reinforced with the halloysite particles. *Materials Science Forum* 783: 1591–1596.

Downing, T.D., Kumar, R., Cross, W.M., Kjerengtroen, L., Kellar, J.J. 2000. Determining the interphase thickness and properties in polymer matrix composites using phase imaging atomic force microscopy and nanoindentation. *Journal of Adhesion Science and Technology* 14: 1801–1812. http://dx.doi.org/10.1163/1568561007432848

Durate, M., Benitez, A., Gomez, K., Zuluaga, B., Meza, J., Cadona-Maya, Y., Rudas, J.S., Isaza, C. 2020. Nanomechanical characterization of a metal matrix composite reinforced with carbon nanotubes. *AIMS Materials Science* 7: 33–45.

Gawdzinska, K., Bryll, K., Guema, M., Pijanowski, M. 2014. Description of surface topography of metal matrix composite castings. *Archieves of Foundary Engineering* 14: 169–174.

Hanada, K., Nishiyama, T., Yoshitake, T., Nagayama, K. 2010. Optical emission spectroscopy of deposition process of ultrananocrystalline diamond/hydrogenated amorphous carbon composite films by using a coaxial arc plasma gun. *Diamond & Related Materials* 19: 899–903. doi: 10.1016/j.diamond.2010.02.019

Hancock, J.R. 1967. Transmission electron microscopy of interfacial areas in metal matrix composites. *Journal of Composite Materials* 1: 136–142.

Hosseini, M.G., Sabouri, M., Shahrabi, T. 2008. Comparison of the corrosion protection of mild steel by polypyrrole–phosphate and polypyrrole–tungstenate coatings. *Journal of Applied Polymer Science* 110: 2733–2741.

Joyce, P.T., Kugler, D., Moon, T.J. 1997. A technique for characterizing process – Induced fiber waviness in unidirectional composite laminates-using optical microscopy. *Journal of Composite Materials* 31: 1694–1727.

Kalaleh, H., Tally, M., Atassi, Y. 2013. Preparation of a clay based superabsorbent polymer composite of copolymer poly (acrylate-co-acrylamide) with bentonite via microwave radiation. *Research & Reviews in Polymer* 4: 145–150.

Kayathri, J., Meiyammai, N.R., Rani, S., Bhuvana, K.P., Palanivelu K., Nayak, S.K. 2013. Study on the optoelectronic properties of uv luminescent polymer: ZnO nanoparticles dispersed PANI. *Journal of Materials* 2013: 1–8. http://dx.doi.org/10.1155/2013/473217

Koleukh, D.S., Kaygorodov, A.S., Zayats, S.V., Paranin, S.N. 2018. Structure study of $Al^+Al_2O_3$ composite by atomic force microscopy. *Eurasian Journal of Physics and Functional Matrials* 2: 326–330.

Krylova, V., Dukštienė, N. 2019. The structure of PA-Se-S-Cd composite materials probed with FTIR spectroscopy. *Applied Surface Science* 470: 462–471. https://doi.org/10.1016/j.apsusc.2018.11.121

Lee, P.W., Avishai N., Pokorski, J.K. 2017. Analysis of polymer-biomacromolecule composites in the solid-state via energy dispersive spectroscopy-scanning electron microscopy. *Microscopy and Microanalysis* 23: 1386–1387. doi: 10.1017/S1431927617007590

Liu, K., Liu, H., Chan, P.K., Liu, T., Pei, S. 2018. Age estimation via fusion of depth wise separable convolutional neural networks, *2018 IEEE International Workshop on Information Forensics and Security (WIFS)*, Hong Kong, pp. 1–8.

Lu, N., Lu, X., Jin X., Lu, C. 2007. Preparation and characterization of uv-curable Zno/polymer nanocomposite films. *Polymer International* 56: 138–143.

Montealegre-Melende, I., Arévalo, C., Ariza, E., Pérez-Soriano, E.M., Rubio-Escudero, C., Kitzmantel M., Neubauer, E. 2017. Analysis of the microstructure and mechanical properties of titanium-based composites reinforced by secondary phases and B_4C particles produced via direct hot pressing. *Materials* 10: 1–12. doi: 10.3390/ma10111240

Montealegre-Melende, I., Arévalo, C., Pérez-Soriano, E.M., Kitzmantel, M., Neubauer, E. 2018. Microstructural and xrd analysis and study of the properties of the system Ti-TiAl-B4C processed under different operational conditions. *Metals* 8: 1–12. doi: 10.3390/met8050367

Mukhopadhyay, S.M. 2003. Sample preparation for microscopic and spectroscopic characterization of solid surfaces and films. In *Sample Preparation Techniques in Analytical Chemistry*, ed. Somenath Mitra, S., 377–411. John Wiley & Sons, Hoboken, New Jersey.

Nakamura, Y. 2018. Nanostructure design for drastic reduction of thermal conductivity while preserving high electrical conductivity. *Science and Technology of Advanced Materials* 19: 31–43. https://doi.org/10.1080/14686996.2017.1413918

Nishchev, K.N., Novopoltsev, M.I., VMishkin, V.P., Shchetanov, B.V. 2013. Studying the microstructure of AlSiC metal matrix composite material by scanning electron microscopy. *Bulletin of the Russian Academy of Sciences, Physics* 8: 981–985. doi:10.3103/S1062873813080315

Novoselov, S., Jiang, D., Schedin, F., Booth, T.J., Khotkevich, V.V., Morozov S, V., Geim, A.K. 2015. Two-dimensional atomic crystals. *Proceeding of the National Academy of Sciences of the United States of America* 102: 10451–10453. https://doi.org/10.1073/pnas.0502848102

Orbulov, I.N., Nemeth, A., Dobranszky, J. 2008. XRD and EDS investigations of metal matrix composites and syntactic foams. *13th European Conference on X-ray Septometry*: Cavtat, Crotia.

Palanivel, A., Veerabathiran, A., Duruvasalu, R. 2017. Dynamic mechanical analysis and crystalline analysis of hemp fiber reinforced cellulose filled epoxy composite. *Polímeros* 27: 309–319. http://dx.doi.org/10.1590/0104-1428.00516

Pastukhov, A.V., Davankov, V.A. 2011. Transmission electron microscopy of composite materials based on hypercrosslinked polystyrenes with nanodispersed inorganic compounds. *Bulletin of the Russian Academy of Sciences. Physics* 75: 1248–1250. doi: 10.3103/S106287381109022X

Patel, A.K., Jain, N., Patel, P., Das, K., Bajpai, R. 2018. Crystalline and absorption studies on PMMA/CdS composite using XRD & UV-vis techniques. *AIP Conference Proceedings*, 1942.

Porras, A., Maranon, A. 2012. Development and characterization of a laminate composite material from polylactic acid (PLA) and woven bamboo fabric. *Composite: Part B* 43: 2782–2788. http://dx.doi.org/10.1016/j.compositesb.2012.04.039

Putri, C.R., Syahrial, A.Z., Yunus, S., Utomo, B.W. 2019. Optimizing the addition of TiB to improve mechanical properties of the ADC 12/SiC composite through stir casting process, *E3S Web of Conferences 130*. https://doi.org/10.1051/e3sconf/201913001005

Qing, W., Min, L., Yizhuo, G., Shaokai, W., Zuoguang, Z. 2015. Imaging the interphase of carbon fiber composites using transmission electron microscopy: Preparations by focused ion beam, ion beam etching, and ultramicrotomy. *Chinese Journal of Aeronautics* 28: 1529–1538. http://dx.doi.org/10.1016/j.cja.2015.05.005

Rein, A., Seelenbinder, J. 2015. Composite thermal damage – correlation of short beam shear data with FTIR spectroscopy, Application Note, Agilent Technologies.

Rydz, J., Siskov, A., Eckstein, A.A. 2019. Scanning electron microscopy and atomic force microscopy: Topographic and dynamical surface studies of blends, composites, and hybrid functional materials for sustainable future. *Advances in Materials Science and Engineering* 2019: 1–15. https://doi.org/10.1155/2019/6871785

Salame, P.H., Pawade, V.B., Bhanvase, B.A. 2018. Characterization tools and techniques for nanomaterials. In *Nanomaterials for Green Energy*, eds. Bhanvase B.A., Pawade V.B., Dhoble S.J., Sonawane S.H., Ashokkumar M., 83–111. Elsevier.

Sauvage, X., Wetscher, F., Pareige, P. 2005. Mechanical alloying of Cu and Fe induced by severe plastic deformation of a Cu-Fe composite. *Acta Materialia* 53: 2127–2135. http://doi.org/10.1016/j.actamat.2005.01.024

Seelenbinder, J. 2015. Measurement of composite surface contamination using the Agilent 4100 ExoScan FTIR with diffuse reflectance sampling interface, Application Note, Agilent Technologies.

Senthilbabu, S., Vinaygam, B.K., Arunraj, J. 2015. Roughness analysis of the holes drilled on Al/SiC metal matrix composites using atomic force microscope. *Applied Mechanics and Materials* 766, 837–843.

Sharp, N.D., Goodsell, J.E., Favaloro, A.J. 2019. Measuring fiber orientation of elliptical fibers from optical microscopy. *Journal of Composites Science* 3: 1–15. doi:10.3390/jcs3010023

Sousa, F.D.B., Scuracchio, C.H. 2014. The use of atomic force microscopy as an important technique to analyze the dispersion of nanometric fillers and morphology in nanocomposites and polymer blends based on elastomers. *Polimeros* 24: 1–12. http://doi.org/10.1590/0104-1428.1648

Srivastava, S., Haridas M., Basu, J.K. 2008. Optical properties of polymer composite. *Bulletin of Materials Science* 31: 213–217.

Stark, N.M., Matuana, L.M. 2007. Characterization of weathered wood-plastic composite surfaces using FTIR spectroscopy, contactangle and XPS. *Polymer Degradation and Stability* 92: 1883–1890. doi: 10.1016/j.polymdegradstab.2007.06.017

Strankowski, M., Wlodarczyk, D., Piszczyk, A., Strankowska, J. 2016. Polyurethane nanocomposites containing reduced graphene oxide, FTIR, Raman, and XRD studies. *Journal of Spectroscopy* 2016: 1–7. http://dx.doi.org/10.1155/2016/7520741

Veklich, A., Boretskij, V., Kleshych, M., Fesenko, S., Murmantsev, A., Ivanisik, A., Khomenko, Tolochyn, O., Bartlova, M. 2019. Thermal plasma of electric arc discharge between Cu–Cr composite electrodes. *Plasma Physics and Technology* 6: 27–30. doi: 10.14311/ppt.2019.1.27

Wadatkar, N.S., Waghuley, S.A. 2012. Characterization of pani-PVAC composite thin films using XRD analysis. *Scientific Reviews & Chemical Communications* 2: 392–394.

Wang, H. 1996. Scanning force microscopy study of ceramic fibers for advanced composites. *Journal of the American Ceramic Society* 79: 2341–2344.

Weiss, P., Lapkowski, M., LeGeros, R.Z., Bouler, J.M., Jean, A., Daculsi, G. 1997. FTIR spectroscopic study of an organic/mineral composite for bone and dental substitute materials. *Journal of Material Science: Materials in Medicine* 8: 621–629.

Yan, S., Jiang, C., Guo, J., Fan, Y., Zhang, Y. 2019. Synthesis of silver nanparticles loaded onto polymer-inorganic composite materials and their regulated catalytic activity. *Polymers* 11: 401–423. doi: 1.3390/polym11030401.

Yen, S.C., Chu, T.C., Jao, H. 1991. Applications of optical microscopic image analysis to composite materials. *Experimental Techniques* 19: 21–23. https://doi.org/10.1111/j.1747-1567.1991.tb01208.x

/ 8 Experimental Evaluation on Tribological Behavior of TiO_2 Reinforced Polyamide Composites Validated by Taguchi and Machine Learning Methods

S. Sathees Kumar and Ch. Nithin Chakravarthy
CMR institute of Technology

R. Muthalagu
B.V. Raju Institute of Technology

CONTENTS

- 8.1 Introduction 170
- 8.2 Polyamides 171
- 8.3 Polyamide 6 171
- 8.4 Chemical Composition of PA6 171
- 8.5 Features of PA6 172
- 8.6 Applications of PA6 173
- 8.7 Fillers 173
- 8.8 Materials Used 173
- 8.9 Preparation of the Test Specimen 173
  - 8.9.1 Preparation of the PA6 Test Specimen 173
  - 8.9.2 Various Stages of Preparation of the PA6 Composite Test Specimen 174
  - 8.9.3 Proportions of the Fabricated PA6 Composite Test Specimen 174
- 8.10 Tribological Properties of PA6 Composites 175
  - 8.10.1 Friction and Wear Rate 175

DOI: 10.1201/9781003121954-8

8.11 Analysis of the Tribological Performance of PA6 Mixtures 176
    8.11.1 Co-Efficient of Friction for PA6 Mixtures................................. 177
    8.11.2 Rate of Wear 179
8.12 SEM Study of TiO_2/PA6 Composites 180
8.13 Comparative Study on Validation.. 182
    8.13.1 Experimental Design 182
    8.13.2 Exploration of Experimental Design Results 182
    8.13.3 ANOVA and Effects of Factors 183
8.14 Machine Learning 185
    8.14.1 Linear Regression 185
    8.14.2 Comparison of Wear Vs Weight% 187
    8.14.3 Comparison of Wear Vs Speed.. 187
    8.14.4 Comparison of Wear Vs Load 188
    8.14.5 Validation of Experiment, Taguchi Method, and ML Algorithm 188
    8.14.6 Comparison of the Taguchi Method and the ML Algorithm............ 188
8.15 Conclusions... 189
References.. 189

8.1 INTRODUCTION

Plastic materials are regularly employed for various purposes, for instance, bearings, seals, and gears, considering their certain inclinations over alloys. Among the most much of the time utilized plastic materials are the polyamides (PA) owing to their mechanical properties (Adams, 1988). When all is said and done, the PA has great wear and abrasive protection, high quality, and strength, which are joined with acceptable effect resistance proposed (Strong & Brent, 2006). The primary weakness of PA is their high moisture retention, which brings down their mechanical properties and dimensional solidness checked on (Ebewele, 2000; Strong & Brent, 2006). Additionally, PA materials lose their mechanical properties at high temperatures and it has been researched (Mao, 2007). To overcome the above limitations, some researchers have reported that the mechanical and wear resistances can be improved, when the polymers are reinforced with fillers which was studied (Sung & Suh, 1979; Chen et al., 2008; Kowandy et al., 2008). The usual behavior of adding fillers with the polymers will increase some properties like mechanical, tribological, and thermal stabilities of the polymer (Unal & Mimaroglu, 2012). Fillers are fundamentally basic inorganic mineral powders, added to improve handling, rigidity, and dimensional constancy as expressed (Brydson, 1966).

    The various reinforcements used in PA materials are graphite, carbon, wax, polytetrafluoroethylene (PTFE), polyethylene terephthalate (PET), silica, carbon nanotube, carbon fiber, and high-density polyethylene (HDPE). In this experimental work, to improve the tribological properties of pure Polyamide 6, it is decided to reinforce with a filler of TiO_2 with different weight proportions. The main influences of reinforcing TiO_2 filler are as follows: as of late, with an end goal to get unique hygienic materials striving with organic and chemical contagion ready, a consideration has been disbursed to TiO_2 owing to their photocatalytic movement

(Kubacka et al., 2008; Serrano et al., 2012). Not many works have been found in the past examination about the impacts of TiO_2 on the properties of thermoplastics. The impacts of TiO_2 on the formation of starch/polycaprolactone mixes showed the development of hydrogen bonds between TiO_2 and the medium. A definite investigation of the impacts of these boundaries on the composite properties will be profoundly enlightening. A solid interfacial bond can adequately move load from the lattice to the polymer and subsequently can improve the general execution of the composite (Fei et al., 2013).

8.2 POLYAMIDES

A point-by-point investigation of the impacts of these boundaries on the composite properties of PA is a significant gathering of the thermoplastic polycondensates. The amide gathering can be obtained by the polymerization of lactams (polylactams) or by the buildup of diamines with dicarboxylic acids (nylons). Hardly any creators expressed that PA continually pull in more extensive intrigue on account of their interesting mechanical, thermal, and morphological properties. PA are notable for their fantastic mechanical performances. The two substantial sorts of PA are polyamide 66 (PA66) and PA6. PA66 is made out of two basic units, each with six carbon atoms, in particular, the residues of hexamethyl endiamine and adipic acid (Hatke et al., 1991; Spiliopoulos & Mikroyannidis, 1998; Liaw et al., 2003).

PA6 or poly 6-caprolactam, another significant polymer, involves a single structural unit, to be specific, the clear residue of 6-aminocaproic acid. PA do uncover an inclination to creep under applied load. Likewise, the properties of PA are extensively influenced by moisture. PA have a few advantages over different classes of polymers. The thermoplastic polymers are a class of engineering materials for developing commercial outcomes, because of their simplicity of assembling and great thermomechanical properties (Benaarbia et al., 2014).

8.3 POLYAMIDE 6

Polyamide 6 (PA6) is a polymer created by Paul schlack at Interessen—Gemeinschoft Farben to reproduce the properties of Polyamide 66 (PA66) without disregarding the patent of its creation. In contrast to most different PA, PA6 is not a buildup polymer and rather, it is encircled by ring-opening polymerization. PA6 begins as crude caprolactam. As caprolactam has six carbon molecules, it has the name Nylon 6. When caprolactam is warmed to about 533 K in an inert environment of nitrogen for around 4–5 hours, the ring breaks and encounters polymerization. By then, the fluid mass has experienced the spinners to shape fibers of Nylon 6.

8.4 CHEMICAL COMPOSITION OF PA6

PA are a gathering of thermoplastic polymers containing amide bunches (–CONH) in the fundamental chain. They are famously known as Nylon 6. PA 6 [NH– $(CH_2)_5$– CO] is produced using ε-caprolactam. It is framed by ring-opening polymerization of ε-caprolactam.

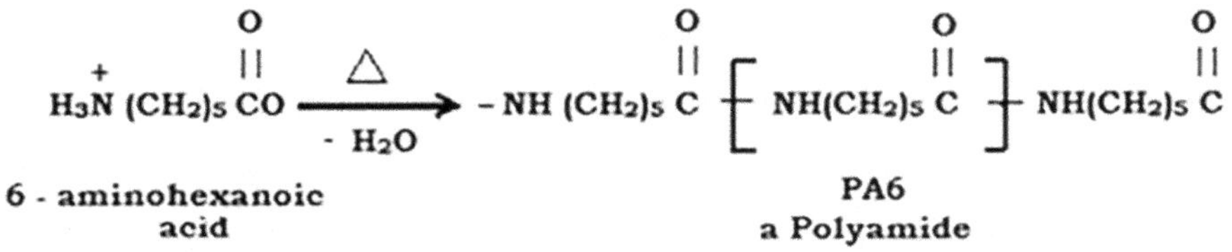

$$H_3N^+(CH_2)_5CO \xrightarrow[-H_2O]{\Delta} -NH(CH_2)_5\overset{O}{\overset{\|}{C}}\left[NH(CH_2)_5\overset{O}{\overset{\|}{C}}\right]NH(CH_2)_5\overset{O}{\overset{\|}{C}}$$

6 - aminohexanoic acid

PA6
a Polyamide

FIGURE 8.1 Chemical composition of PA6. (*Source:* S. Sathees Kumar & G. Kanagaraj, *Bulletin of Material Science*, 2016, 1467–1481.)

"The chemical composition of PA6 is shown in Figure 8.1".
Alternative name: poly-ε-caproamide
Trade names: Capron, Ultramid, Nylatron
Class: Aliphatic polyamides

8.5 FEATURES OF PA6

PA6 strands are intense and have high rigidity, just as versatility and radiance. They are of wrinkle conformation. They are additionally resistant to abrasion and synthetic elements, for instance, acids and solvable bases (Pogacnik & Kalin, 2012). The physical properties of PA6 are acceptable wear and abrasive protections, low coefficient of friction, high flexibility, high quality, and durability joined with great effect resistance. PA6 is a crystalline thermoplastic that has a low thermal coefficient with linear expansion, low coefficient with thermal development, and exceptional sensitivity to dampness. PA6 has become a strong competitor network, attributable to its great thermal stability, low dielectric consistency, and high elasticity (Karsli & Aytac, 2013). The general features of PA6 are given in Table 8.1.

TABLE 8.1
General Features of PA6

Property	Value
Glass transition temperature	40°C
Melting temperature	220°C
Density (crystalline) at 25°C	1.23 g/cc
Density (amorphous) at 25°C	1.084 g/cc
Tensile strength	35–230 MPa
Tensile modulus	2000–3000 MPa
Flexural strength	40–230 MPa
Flexural modulus	1800–2414 MPa
Heat deflection temperature	58°C
Surface hardness (Shore D)	79

Source: Data from designerdata.nl/materials/plastics/thermo-plastics/polyamide-6.

8.6 APPLICATIONS OF PA6

PA are prospective thermoplastic materials used for innumerable purposes, inferable from their amazing expansive properties, which were recognized (Li et al., 2013). PA6 is utilized as a string in bristles for toothbrushes, surgical stitches, strings for acoustic and traditional instruments, including guitars, violins, violas, and cellos. It is likewise utilized in the production of an enormous assortment of strings, ropes, fibers, nets, and tire ropes, just as hosiery and weaved articles of clothing. In the automobile industry, it is employed for wire and link jacketing, cooling fans, air admission, turbo air channels, valve and engine covers, brake and force guiding repositories, gears for windshield wipers, speedometers, and numerous other automotive parts (Mallick, 2007). It is utilized as gear and bearing materials as a result of their equalization in quality, hardness, and sturdiness and in view of their great friction characters (Bermudez et al., 2001).

8.7 FILLERS

Generally, fillers are considered as added substances; in view of their unfavorable geometrical characters and surface area or surface chemical composition, they can just moderately update the modulus of the polymer, while their quality remains unaltered or even decreased. Particulate filler reinforced mixtures have various recompenses around neat resin media, incorporating heightened stiffness, strength and dimensional dependable qualities, upgraded durability or impact, improved heat distortion, expanded mechanical damping, diminished penetrability to gases and fluids, adjusted electrical properties, and degraded prices (Nielsen & Landel, 1994; Kumar & Wang, 1997). The particulate filler for polymer composite systems are open in a couple of sizes and shapes including sphere, cubic, platelet, or some other predictable or uneven geometry (Katz, 1998). With the expansion of filler other than the mechanical properties, many different attributes of the material can be improved (Konieczny et al., 2013).

8.8 MATERIALS USED

PA6: Nylon 6, (coded as PA6) 3 mm pellet size, density—1.40 g/cm^3, physical state—white, and appearance—chips. Similarly, titanium dioxide, then termed titanium (IV) oxide or titania, is the typically proceeding oxide of titanium, with the chemical formula TiO_2. When used as a shade, it is called titanium white and pigment white. Pigment grade TiO_2 has a median particle size in the range of 200–300 nm and a density of 4.23 g/cm^3. PA6 and TiO_2 fillers are shown in Figure 8.2.

8.9 PREPARATION OF THE TEST SPECIMEN

8.9.1 Preparation of the PA6 Test Specimen

The melting temperature of PA6 is 220°C. As a result, PA6 pills are filled in the injection molding equipment and they are directly molten by the injection molding device. At this hotness (210°C), PA6 is melted and converted into a molten state. The obtained liquid PA6 is passed from injection molding to a preheated die. The preheated die

FIGURE 8.2 PA6 and fillers: (a) PA6 and (b) TiO_2.

is used to fabricate the specimens as per the following dimensions: 35 mm length and 25 mm diameter for tribological experimental tests. The obtained specimens are machined as per the ASTM test standards for tribological tests.

8.9.2 Various Stages of Preparation of the PA6 Composite Test Specimen

Stage 1: Preheating

Fillers (TiO_2) have high melting focuses and subsequently, they will not blend with PA6 promptly. Hence, fillers should be preheated under the liquifying hotness. Here, the fillers are warmed under the liquifying hotness right throughout the muffle furnace up to 1740°C.

Stage 2: Cooling

As a rule, the liquifying hotness of PA6 is 210°C. If it is rapidly mixed with a filler, it will fire. Subsequently, to reach the hotness of 1740°C, the filler was ventilated at air heat for 40–50 minutes. By this cooling procedure, the filler has transformed into ash formation.

Stage 3: Stirring/Blending

The cooled filler was mixed with PA6 pills by a blending process. The blending procedure is done in a 75 mm diameter and 100 mm height cast iron crucible.

Stage 4: Molding

The mixed material is liquefied with the hotness up to 210°C in the injection molding device. In this heat, PA6 liquifies and arrives at the molten state. The melted PA6 is delivered from the injection molding device to a 75°C warmed die. Now, the prewarming of the die is useful for a smooth stream of mixture materials. The fabrication of specimen process and fabricated specimens are shown in Figures 8.3 and 8.4, respectively.

8.9.3 Proportions of the Fabricated PA6 Composite Test Specimen

PA6 95 wt.% with filler 5% wt., PA6 90 wt.% with filler 10 wt.%, PA6 80% wt. with filler 20 wt.%, and PA6 70 wt.% with filler 30% wt. Temperature varieties in each phase of sample preparation and particulars of manufactured test samples of PA6 and PA6 composites are listed in Tables 8.2 and 8.3, respectively.

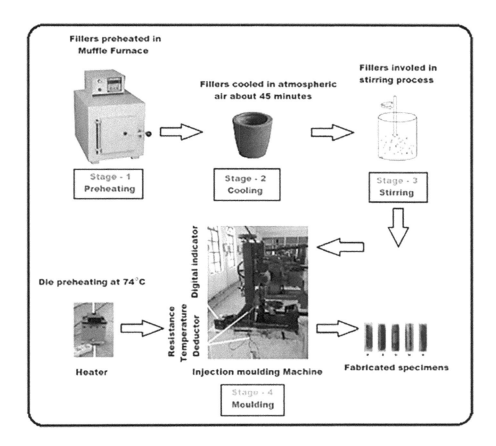

FIGURE 8.3 Fabrication process of fillers reinforced PA6 test specimens.

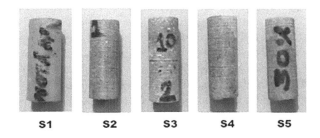

FIGURE 8.4 Fabricated PA6 and PA6 composite specimens.

8.10 TRIBOLOGICAL PROPERTIES OF PA6 COMPOSITES

8.10.1 Friction and Wear Rate

The dry gliding wear qualities of the mixture were evaluated using a pin-on-plate analyser as indicated by the ASTM G99 standard. The size of the test is 12 mm width and 30 mm length. The exteriors of both the example and the plate were

TABLE 8.2
Details of PA6 and PA6 Composite Test Specimens

Specimen Code	PA6 (wt.(%))	TiO$_2$ (wt.(%))	Remarks
S1	100	–	Raw PA6
S2	95	5	TiO$_2$-reinforced PA6 specimen
S3	90	10	
S4	80	20	
S5	70	30	

TABLE 8.3
Temperature Variations in Each Stage of Specimen Preparation

Test Specimen	Stage 1 Preheating Temp. of Fillers (°C)	Stage 2 Cooling Temp. (°C)	Stage 3 Stirring Temp. of PA6 (°C)	Stage 4 Molding Temp. of PA6 (°C)	Code No of Test Specimen
PA6	–	31	–	210	S1
TiO$_2$/PA6 specimen	1740	31	122–128	210	S2–S5

cleansed with a delicate paper absorbed acetone prior to the test. The testing rate is possessed in the span of 1000, 1500, and 2000 rpm. At that point, the load is attached to the scope of 5, 10, 20, and 30 N. The frictional coefficient and wear features of five samples are tried in the pin-on-disc/plate tester under various abrasion settings. A tribometer formats since the contact happening between two-level surfaces prevents a huge difference in the region of contact between the polymer and the harder counter face because of wear and creep as researched (Palabiyik & Bahadur, 2002). The atmospheric circumstance in the test center is 30°C and 46% humidity. The basic and preceding loads of the specimens were assessed by exploiting an electronic digital balance. The division between the basic and remaining loads is the proportion of the weight discount. The measure of wear is resolved as far as weight reduction by weighing samples during the test. Wear results are normally gained by operating a test on chosen sliding distance and chosen estimations of load and speed. A few papers are tended to as of now the wear execution of modified thermoplastics and thermosets (Sung & Suh, 1979). Here, the accompanying requirements and the trail specifications as provided in Table 8.4 are used as a portion of the tribometer.

8.11 ANALYSIS OF THE TRIBOLOGICAL PERFORMANCE OF PA6 MIXTURES

From the trial consequences of tribological tests, the coefficient of contact and wear rate results are attained. Tribological execution of PA6 mixtures are shown in Table 8.5.

TABLE 8.4
Tribological Testing Specifications of TiO$_2$-Strengthened PA6 Composites

Constraints	Testing Conditions
Applied load	5, 10, 20 and 30 N
Material of disc	EN31 steel
Speed of disc	1000–2000 rpm
Materials of pin	TiO$_2$ reinforced PA6 (S2–S5)
Surface condition	Dry
Diameter of track	100 mm
Distance of sliding	1000 m

TABLE 8.5
Tribological Execution of PA6 and PA6 Composites at Different Loads

Specimen Code	Friction Co-efficient with Loads				Wear Rate×10^{-3} (mm^3/Nm) with Loads			
	5 N	10 N	20 N	30 N	5 N	10 N	20 N	30 N
S1	0.32	0.35	0.37	0.42	7.2	7.5	7.8	8.2
S2	0.29	0.33	0.34	0.35	6.6	6.8	7.0	7.3
S3	0.30	0.34	0.35	0.38	6.8	7.0	7.1	7.4
S4	0.32	0.35	0.36	0.39	6.2	6.4	6.7	6.9
S5	0.27	0.31	0.32	0.34	6.9	7.1	7.3	7.5

8.11.1 Co-Efficient of Friction for PA6 Mixtures

Figure 8.5a–d exposes the varieties of frictional coefficient for PA6 (S1) and TiO$_2$-strengthened PA6 mixtures (S2–S4) for various loads. The drop in frictional coefficient with an expansion in weight is ascribed as a result of the greatest sliding speeds at various weight percentages of reinforcement. The TiO$_2$-strengthened PA6 composite materials have an inferior coefficient of friction contrasted with raw PA6. Be that as it may, it is seen through the analysis that the composite material illustrates a trivial variation in the frictional coefficient with the variabilities of load and sliding speed. In Figure 8.5, TiO$_2$-strengthened PA6 composite materials have a lesser coefficient friction contrasted with raw PA6. From the outcomes, it is distinguished that the PA6 with 30 wt.% (S5) has the least frictional coefficient at all loads. It is presumed that the expansion of load up demonstrates a substantial increment in the frictional coefficient as shown in Figure 8.5a–d. Taking all these outcomes, it is comprehended that the occurrence of TiO$_2$ performs to create consequences for the frictional coefficient of PA6. Clearly, the frictional coefficient builds at first to a higher value, because of the new rough material, and as the procedure proceeds, it almost stays the same for the whole test. It is likewise noticed that the coefficient of friction diminishes when the filler substance increments. It is exposed that the estimation of frictional co-efficient is quite at primary weights in raw PA6 and fortified PA6 composites.

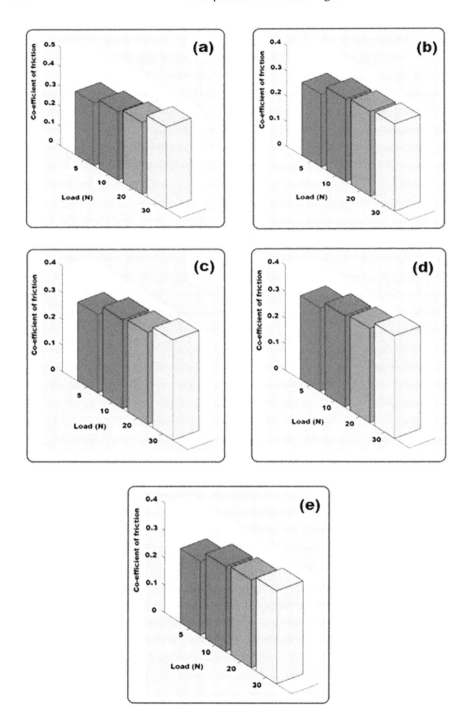

FIGURE 8.5 Co-efficient of friction for (a) S1, (b) S2, (c) S3, (d) S4, and (e) S5 composites.

8.11.2 Rate of Wear

The wear rate is predicted to make use of the given formula:

$$\text{Wear rate} = (m/\rho L d) \times 10^3 \text{ mm}^3/\text{Nm}. \tag{8.1}$$

where m is the loss of mass in grams, ρ is the test material density in g/cm³, L is the applied load in Newton, and d is the distance of sliding in meters. The readings of the diagram are plotted at a normal time interval of 5 minutes. The sample surface is estimated utilizing an automatic balance with the precision of ±0.01 mg.

During the primary trials, the exteriors of both the samples and the steel matching parts are coarse and accordingly solid intercorrelations involving the outsides bring about a high frictional coefficient. Such as the wear procedure proceeds, the coarse contours of the steel matching part and the samples are softened. With an expansion in the sliding speed, the frictional hotness is significantly high, and the extra deterioration fragments follow on the sample. Figure 8.6 illustrates the wear attributes of PA6 and TiO$_2$-reinforced PA6 mixtures in 5, 10, 20, and 30 N loads. Applied weight is one of the few highly considerable variables influencing the deterioration of the composites. The deterioration rate of the composite material has critical variations contrasted with the raw PA6 material at entire loads. In Figure 8.6, TiO$_2$-filled PA6 composites (S2–S5) show remarkable weakening in wear rate. In Figure 8.6a, the particular wear rate is moderately high at the load of 5 N. It happens due to the less quantity of fillers. In the tribological test, though the sample interacts with the disc, wear is created on the interaction exterior of the sample. The abrasion wear resistance is significantly expanded at higher loads in light of the fact that the greater part of the filler particles infiltrate into the PA6 polymer surface and furthermore ensures the wear debris in the wear test. The aftereffect of Figure 8.6 shows that the PA6-reinforced 30 wt.% TiO$_2$ (S5) has a diminished wear rate clearly contrasted with raw PA6 and different composites. TiO$_2$ particles firmly bond with the primary material. They ensure the outside opposed to extreme destructive activity in the counterface. The wear of the composites lessens with increment in the

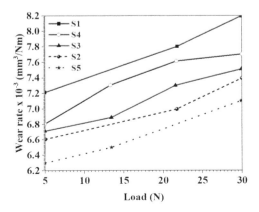

FIGURE 8.6 Wear rate of PA6 and PA6 composites.

level of TiO_2 particles as signified in Figure 8.5; however, the frictional coefficient of the composites weakens with an expansion in the substance of TiO_2 particles as determined in Figure 8.6. Figure 8.6 illustrates that the wear rate in the limit of 7.4×10^{-3} mm^3/Nm and 6.6×10^{-3} mm^3/Nm for PA6 with 5 wt.% TiO_2 composite, 7.5×10^{-3} mm^3/Nm, and 6.7×10^{-3} mm^3/Nm for PA6 with 10 wt.% TiO_2 mixture and the deterioration rates are in the scope of 7.7×10^{-3} mm^3/Nm and 6.8×10^{-3} mm^3/Nm for PA6 with 20 wt.% TiO_2 composite.

Being the last fact, the normal wear rate is the limit of 7.1×10^{-3} mm^3/Nm and 6.3×10^{-3} mm^3/Nm for PA6 with 30 wt.% TiO_2 composites. Be that as it may, the normal wear rate is 8.2×10^{-3} mm^3/Nm and 7.5×10^{-3} mm^3/Nm for raw PA6. Subsequently, PA6 with 30 wt.% TiO_2 mixture produced wear obstruction and quite frictional coefficient on an exterior level contrasted and PA6 and different compositions. Generally, the TiO_2 particles have inside and out upgraded hardness and wear opposition. In a similar manner, PA6 reinforced 30 wt.%. TiO_2 had a quite frictional coefficient and a discreet wear rate of 7.1×10^{-3} mm^3/Nm. Regardless, the result of Figure 8.6. shows that the PA6 reinforced 30 wt.% TiO_2 wear rate was lessened, as clearly observed by analysing PA6 and various mixtures. In the tribological test, the sample associates with the plate/disc and the basic attributes of warmth and wear have been made on the interaction of an exterior of the sample. TiO_2 elements have high stiffness and superior thermal conductivity. In this way, it has fortified with PA6, and the attributes of PA6 have been improved. Now, the elements dissipated on the PA6 test interact with the disc. The hotness and wear has covered from the interaction exterior of the tribometer. Thus, it has created extreme wear opposition on the surface of the example.

8.12 SEM STUDY OF TIO_2/PA6 COMPOSITES

Figure 8.7a and b reveals the morphology depictions for a raw PA6 test sample of when worn sides. Few destroyed elements and plastic distortion were discovered before sported surfaces of raw PA6 in Figure 8.7a. In Figure 8.7b, the wedge course of action and plastic bending were made on the faces, additional voids, and little extension parts were discovered after the worn surfaces of PA6. Furthermore in Figure 8.7c and Figure 8.7d TiO_2 particles were dependably scattered in PA6 . Then PA6 with 5 wt.%. TiO_2 interprets the plastic distortion and voids were made less differed and PA6 from 5 wt.% TiO_2. Broken particles, little scope breaks, and wedge headway were not uncovered outwardly of PA6 with 5 wt.%. TiO_2. Figure 8.7e shows the voids strategy and damaged elements of PA6 with 10 wt.%. TiO_2. Figure 8.7f shows the gaps growth, plastic mutilation, and damaged particles of PA6 with 10 wt.% TiO_2. It is clarified that TiO_2 elements were not dependably scattered in the PA6. Figure 8.7g and 8.7h shows the depiction of 20 wt.%. TiO_2 this portrait obviously delineates the identical spread of TiO_2 elements in PA6. In the interim plastic bending and damaged particles were surrounded upon the before worn surfaces. After worn surfaces of 20 wt.%. TiO_2 unreveled miniaturized scale breaks game plan on the exteriors of PA6 (Figure 8.7h). Figure 8.7i shows the gaps improvement, flexible bending, and damaged elements of PA6 with 30 wt.% TiO_2. Figure 8.7j shows the springiness

Evaluation on Tribological Behavior

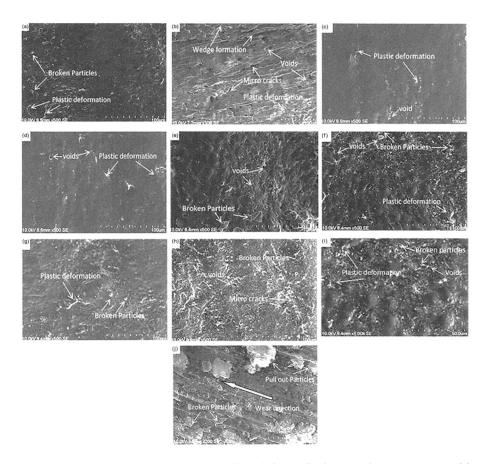

FIGURE 8.7 SEM images of PA6 and TiO$_2$-reinforced PA6 composites (a, c, e, g, and i before wear and b, d, f, h, and j after wear).

considering the way that following wear of the PA6 exterior, the holding of TiO$_2$ elements are not correspondingly spreading at different places on top. These pictures disclose the extension of TiO$_2$ with PA6 and the permeability of the TiO$_2$ material in the PA6. Wear bearing is also apparent in Figure 8.7j (Sathees Kumar & Kanagaraj, 2016a–d, 2017). It shows that better wear happened on the PA6 with 30 wt.% TiO$_2$. The all around utilized appearances of the composite was normally level and there had every one of the stores of being continuously joined wear destruction organized along the sliding track .

It is seen that TiO$_2$ strengthened with PA6 surfaces as a general rule wear even more quickly. The more extent of penetrable of TiO$_2$ is mixed in with PA6 as moreover superior as past arrangement. It is perceived that the dispersing state of TiO$_2$ in the PA6 medium is sensibly consistent in the 30% composites. In summation, the even flow of the TiO$_2$ elements in the micro arrange of the PA6 is the foremost careful influence for the upgrading in the friction and wear attributes.

8.13 COMPARATIVE STUDY ON VALIDATION

8.13.1 Experimental Design

The Taguchi procedure is a normally used handle-on approach for propelling structure boundaries. The procedure is at first proposed as a technique for improving the nature of things using quantifiable and building ideas. Since test approaches are commonly over the top and grim, the need to fulfill the outline destinations with an immaterial number of tests is unmistakably essential. Table 8.2 demonstrates the parts to be centered around and the task of the relating levels. The bunch picked was the L16 (45) which has 16 columns standing out from the number of tests with five segments at three levels. The test configuration is given in Table 8.6.

The Taguchi factorial investigation methodology diminishes it to just 16 runs contributing an uncommon ideal situation in the phase of preliminary period and price. The arrangement of the analyses is as per the following: the principal segment is allotted filler content (wt.%) (A), the subsequent segment to speed (B), and the third section is doled out to load (C).

8.13.2 Exploration of Experimental Design Results

The preliminary information for wear rate is given in Table 8.7. The information declared is customary to two replications. The examinations of the initial information are passed on utilizing the thing MINITAB exceptionally utilized for the course of action of preliminary applications. Prior to taking a gander at the test, information utilizing this approach for expecting the extent of execution, the conceivable joint endeavors between control portions are thought of. This factorial arrangement for the closeness of the collaboration impacts.

Figure 8.8 shows explicitly the influence of the three control segments of definite wear rate of the samples C_1 (5 wt.%), C_2 (10 wt.%), C_3 (20 wt.%), and C_4 (30 wt.%). The study results give the mix factors accomplishing the least obvious wear rate of composites. Assessment of these outcomes prompts the end that segments blend accomplished least explicit wear rate as 30% TiO_2 (C4) content in PA6 as appeared in Figures 8.9 and 8.10.

TABLE 8.6
Stages of Variables Used in the Trials

Parameters	Stage				Units
	I	II	III	IV	
Filler content in wt.	5	10	20	30	%
Speed	500	1000	1500	2000	rpm
Load	5	10	20	30	N

TABLE 8.7
Experimental Design Using L16 Array

Filler in wt.(%)	Speed (rpm)	Load (N)	Wear Rate (mm³/Nm)
5	500	5	6.3
5	1000	10	8.1
5	1500	20	8.5
5	2000	30	8.8
10	500	10	6.9
10	1000	5	6.7
10	1500	30	8.1
10	2000	20	7.8
20	500	20	7
20	1000	30	7.5
20	1500	5	6.5
20	2000	10	7.3
30	500	30	6.1
30	1000	20	6.8
30	1500	10	6.5
30	2000	5	6.4

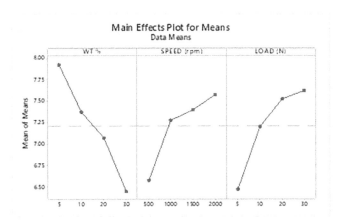

FIGURE 8.8 Plot of outcomes for TiO_2/PA6 composites.

8.13.3 ANOVA AND EFFECTS OF FACTORS

Table 8.8 shows the particular wear rate it is seen that the weight % (21.67), speed (11.02), and load (15.51) have remarkable impacts on the certain wear rate, and in this manner, these are genuinely and verifiably significantly enormous. From the assessment of ANOVA and retort of the S/N extent for express wear rate, it is seen that

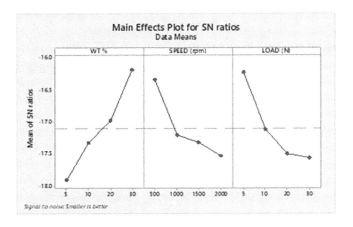

FIGURE 8.9 Plot of influences for S/N ratios of TiO_2/PA6 composites.

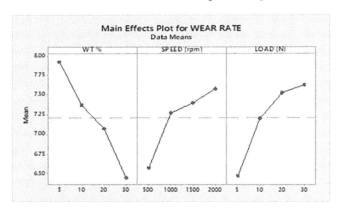

FIGURE 8.10 Plot of impacts for certain wear rate of TiO_2/PA6 composites.

TABLE 8.8
ANOVA of S/N for the Exclusive Wear Rate

Parameters	DOF	Seq SS	Adj MS	F	P
Weight %	3	4.5369	1.51229	21.67	0.001
Speed	3	2.3069	0.76896	11.02	0.007
Load	3	3.2469	1.08229	15.51	0.003
Error	6	0.4188	0.06979		
Total	15	10.5094			

the filler substance in weight % has a huge impact on exact wear amount followed by weight and speed. The TiO_2 fillers with PA6 seemed, by all accounts, to be logically convincing in the decline of the specific wear rate under both 20 and 30 N of applied weight. It might be contemplated that the 30% TiO_2 (C4) fortified with PA6

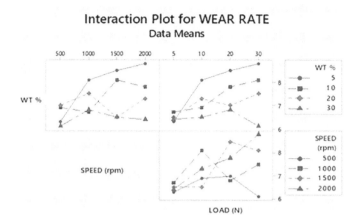

FIGURE 8.11 Interactions plot for specific wear rate.

composites have better tribological properties. The communication plot for explicit wear rate as appeared in Figure 8.11.

From the Taguchi validation results, it is clear that S5 composites have excellent wear resistance properties compared with other specimens (S1, S2, S3, and S4).

8.14 MACHINE LEARNING

Machine learning (ML), a part of Artificial Intelligence (AI), manipulates a grouping of quantifiable and probabilistic procedures that permit workstations to acquire for reality and perceive concealed examples (the associations among's input and output factors) from tremendous and much of the time uproarious datasets (Blum & Langley, 1997; Botu & Ramprasad, 2015; Kononenko & Kukar, 2007; Michalski et al., 2013).

Throughout assessing only, a part of the expected information, ML algorithms can recognize shrouded designs in the information and get acquainted with a focus limit that best aides enter components to an exit boundary (or yield factors) a system alluded to while the arrangement system. Besides the isolated models, desires for hid data centers can be made and mull over theories with a limited proportion of data rather than using a thorough method to manage research at each and every comprehensible datum centers. Starting late, ML has turned our regular day-to-day existence to be progressively invaluable from various perspectives by affecting image recognition, independent driving, email spam location, among others (Butler et al., 2018; Hansen et al., 2013).

8.14.1 Linear Regression

Diverse ML calculations have been delivered for distinctive sorts of acquiring principles, for instance, administered learning, semi-directed learning, and unaided learning. Directed learning is the utmost extensively exploited learning tactic in logical and designing domains. Among altogether directed learning procedures, linear regression is the best principal unique and has existed thought of and use up generally. As opposed to other continuously increasingly muddled (deep) ML models which are every now and again being measured as "black box" paradigms as of their complication, a linear standard has better accountability of the correlation among key in and output factors. The proposition of regression linear is:

$$y = w^T x \qquad (8.2)$$

where w signifies the learnable boundaries or loads (including a bias w_0), x signifies the input factors (tallying a predictable x_0 for the inclination expression), and y represents the exit boundary (subordinate) of the paradigm.

Unique essential procedure to survey the misfortune in a regression paradigm is utilizing the mean squared error (MSE), which is utilized to evaluate in what way nearby or far the wants areas of the certifiable entireties. Right when the blunder limit of a model is portrayed, the heaps of the model can be directed by an improvement calculation, for example, the ordinary stochastic slant plunge or the Adam advancement calculation. Regression has been generally engaged to different evaluation issues due to its straightforwardness and high accountability. For the functions in mixture materials, the linear regression can be referred to forecast the compact nature of warmth pampered wood-based composites (Tiryaki & Aydın, 2014). The regression to guess actual compact quality, where the data factors unite likely assessed strong attributes and blend game plans (Khademi et al., 2017; Young et al., 2019). In those assessments, despite the direct relapse, the creators besides performed dynamically complicated ML paradigms, for example, NN paradigms for their backslide undertakings. The inventors indicated that powerfully confusing ML paradigms, when everything is said and done, offered continuously precise wants. In any case, in those assessments, the direct models gave basic data like which data factors were logically significant for the craving (compressive quality). Detect that regression recognizes an immediate correlation among input factors and the gauge. Utmost issues have various grades of non-linear attributes and straight isn't fitting for remarkably non-linear issues in which the relationship between the data and yield factors cannot be approximated by a direct capacity. On account of good characteristics of direct relapse model of ML, it will in general be used to envision the lead of composites. From these attributes, the linear regression paradigm is used to recognize the wear properties of PA6 and PA6 composites by applying various loads and speeds. Results of linear regression method are shown in Table 8.9.

TABLE 8.9
Results of the Linear Regression Method

Parameter	RMSE	R^2	MSE	MAE
Weight %	0.2809	0.89	0.07891	0.216
Speed	0.3132	0.85	0.09809	0.270
Load	0.2909	0.89	0.08462	0.233

8.14.2 Comparison of Wear Vs Weight%

Figure 8.12 represents the wear of PA6 and PA6 samples Vs weight % of fillers. Here, the true and predicted points are shown blue and light orange colures respectively. The x-axis represents the input variables of weight % and the y-axis signifies the wear of composites. In this Figure 8.12, the predicted points from linear regression are very nearer to true points. Probably more than 85% of predicted values revealed a better relationship with true values. From Table 8.9, RMSE and R^2 values are 0.2809 and 0.89, respectively; these outputs signify the predicted wear rate better interlinking with true values.

8.14.3 Comparison of Wear Vs Speed

Figure 8.13 represents the wear of PA6 and PA6 samples vs speed of disc. In Figure 8.13, the predicted points from linear regression very adjacent to true points. Probably more than 90% of predicted values revealed a better relationship with true values.

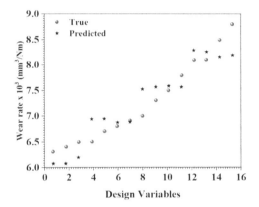

FIGURE 8.12 Prediction of wear rate vs weight %.

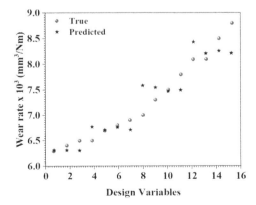

FIGURE 8.13 Prediction of wear rate vs speed.

From Table 8.9, RMSE and R^2 values are 0.3132 and 0.85, respectively; these outputs signify the predicted wear rate better interconnecting with true values.

8.14.4 Comparison of Wear Vs Load

Figure 8.14 represents the wear of PA6 and PA6 samples Vs load applied. In Figure 8.14, the predicted points from linear regression are very close to true points. Probably more than 95% of predicted values revealed a better relationship with true values. From Table 8.9, RMSE and R^2 values are 0.2909 and 0.89 respectively; these outputs signify the predicted wear rate better interconnecting with true values.

8.14.5 Validation of Experiment, Taguchi Method, and ML Algorithm

The validation of experimental, Taguchi, and ML (linear regression) wear rate results of TiO_2-reinforced PA6 composites was carried out. Only high wear resistance obtained, S4 composite specimen was involved in comparisons from experimental, Taguchi, and linear regression methods. Based on the validation, the wear rates (mm³/Nm) of S4 70wt.% PA6 + 30wt.% TiO_2. (S4) specimen attained from the experimental, Taguchi, and linear regression methods are 6.2, 5.5, and 5.8 mm³/Nm, respectively.

8.14.6 Comparison of the Taguchi Method and the ML Algorithm

From the above Taguchi (Table 8.8) and ML methods of linear regression results (Table 8.9), mean square values of the Taguchi method exposed for the weight %, speed, and loads are 1.51229, 0.76896, and 1.08229 respectively. From linear regression, the MSE values for the weight %, speed, and loads are 0.07891, 0.09809, and 0.08462, respectively. However, Figures 8.8–8.10 and 8.12–8.14 illustrate the Taguchi and ML study results, respectively. Figures 8.12–8.14 represent the comparison of theoretical and experimental values. Theoretical values are very nearby to experimental work. Similarly, Figures 8.8–8.12 reveal the Taguchi method for

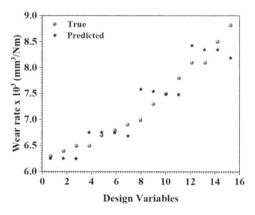

FIGURE 8.14 Prediction of wear rate vs load.

weight, speed, and loads. But the values of the three parameters are just away from the experimental results. In summary, in comparison between the two results, the ML paradigm exposes nearly the accuracy of experimental results compared with the Taguchi method.

8.15 CONCLUSIONS

In this investigation, PA6 mixed with TiO_2 composites was set up by an injection molding device. The impact of TiO_2 substance on the tribological behavior of the PA6 material was assessed at atmospheric temperature. From the Taguchi method and exploratory outcomes, the following conclusions were drawn:

- The impacts of TiO_2 particles on the wear conduct of PA6 polymers were examined. The 30 wt.% TiO_2 particles were ready to progress the wear obstruction of the PA6 polymer composites. The TiO_2 particles appeared to be more efficient in rising the wear opposition when contrasted with the PA6 polymer.
- From the experimental work, the friction co-efficient downsized to 21.42% for PA6 + 30 wt.% TiO_2 composites contrasted and PA6.
- The addition of 30 wt.% measure of TiO_2 diminished the frictional coefficient behavior and enhanced the wear opposition of the PA6 matrix. The wear opposition significantly upgraded up to 22.82% for PA6 + 30 wt.% TiO_2 composites related to PA6.
- The design of studies approach by the Taguchi method and the ML algorithm of linear regression empowered effectively the analyses of the wear conduct of the composites with speed and load as test factors.
- Validated from experimental results, the Taguchi and linear regression method (S4) specimens attained the wear rate results of about 6.2×10^{-3}, 5.5×10^{-3}, and 5.8×10^{-3} mm^3/Nm, respectively.
- The PA6 composites tested in this study may find useful applications in the fabrication of plastic gears for lightly loaded drives with a high level of wear resistance and durability. They transmit power quietly and often without lubrication in applications such as food processors, windshield wiper drives, and even watches.

REFERENCES

Adams, C.C. *Plastic Gearing, Selection and Application*. M. Dekker, New York, 1988.
Benaarbia, A., Chrysochoos, A., and Robert, G. 2014. Kinetics of stored and dissipated energies associated with cyclic loadings of dry polyamide 6.6 specimens. *Polymer Testing* 34: 155–167.
Bermudez, M.D., Carrion, Vilches, F.J., Martínez Mateo, I., and Martínez Nicolas, G. 2001. Comparative study of the tribological properties of polyamide 6 filled with molybdenum disulfide and liquid crystalline additives. *Journal of Applied Polymer Science* 81(10): 2426–2432.
Blum, A.L., and Langley, P. 1997. Selection of relevant features and examples in machine learning. *Artificial Intelligence* 97: 245.

Botu, V., and Ramprasad, R. 2015. Adaptive machine learning framework to accelerate ab initio molecular dynamics. *International Journal of Quantum Chemistry* 115: 1074.

Brydson, J.A. Plastics Materials Princeton, 1966: 39–44, Van Nostrand Co., Inc., Princeton, New Jersey.

Butler, K.T., Davies, D.W., Cartwright, H., Isayev, H., and Walsh, A. 2018. Machine learning for molecular and materials science. *Nature* 559: 547.

Chen, Z., Mo, X., He, C., and Wang, H. 2008. Intermolecular interactions in electrospun collagen–chitosan complex nanofibers. *Carbohydrate Polymers* 72(3): 410–418.

Ebewele, R.O. Polymer *Science* and *Technology*. CRC Press, New York, 2000.

Fei, P., Shi, Y., Zhou, M., Cai, J., Tang, S., and Xiong, H. 2013. Effects of nano-TiO_2 on the properties and structures of starch/poly(-caprolactone) composites. *Journal of Applied Polymer Science*: 4129–4136.

Hansen, K., Montavon, G., Biegler, F., Fazli, S., Rupp, M., Scheffler, M., Von Lilienfeld, O.A., Tkatchenko, A., and Müller, K.R. 2013. Assessment and validation of machine learning methods for predicting molecular atomization energies. *Journal Chemical Theory and Comput*ation 9: 3404.

Hatke, W., Schmidt, H.W., and Heitz, W. 1991. Substituted rod like aromatic polyamides: Synthesis and structure property relations. *Journal of Polymer Science Part A: Polymer Chemistry* 29(10): 1387–1398.

Karsli, N.G., and Aytac, A. 2013. Tensile and thermomechanical properties of short carbon fiber reinforced polyamide 6 composites. *Composites Part B: Engineering* 51: 270–275.

Katz, H.S. Particulate fillers. In S.T. Peters (ed.), Handbook of *Composites*. Springer, Boston, MA, 1998.

Khademi, F., Akbari, M., Jamal, S.M., and Nikoo, M. 2017. Multiple linear regression, artificial neural network, and fuzzy logic prediction of 28 days compressive strength of concrete. *Frontiers of Structural and Civil Engineering* 11: 90.

Konieczny, J., Chmielnicki, B., and Tomiczek, A. 2013. Evaluation of selected properties of PA6-copper/graphite composite. *Journal of Achievements in Materials and Manufacturing Engineering* 60(1): 23–30.

Kononenko, I., and Kukar, M. *Machine Learning and Data Mining: Introduction to Principles and Algorithms*. Horwood Publishing, Chichester, UK, 2007.

Kowandy, C., Richard, C., and Chen, Y.M. 2008. Characterization of wear particles for comprehension of wear mechanisms: Case of PTFE against cast iron. *Wear* 265(11): 1714–1719.

Kubacka, A., Cerrada, M.L., Serrano, M.C., Fernández-García, M., Ferrer, and Fernández-García, M. 2008. Light- driven novel properties of TiO_2-modified polypropylene-based nanocomposite films. *Journal of Nanoscience and Nanotechnology* 8: 3241–3246.

Kumar, S., and Wang Y. 1997. Fibers, fabrics, and fillers. In Composites *Engineering Handbook* (A 98-11526 01-24). Marcel Dekker Inc., New York, vol 11, 51–100.

Li, Z., Xiao, P., Xiong, X., and Huang, B.Y. 2013. Preparation and tribological properties of C fiber reinforced C/SiC dual matrix composites fabrication by liquid silicon infiltration. Solid State Sciences l(16): 6–12.

Liaw, D.J., Chen, W.H., and Huang, C.C. *Polyimides and Other High Temperature Polymers*, Mittal, K.L., ed., 47–70, 2003.

Mallick, P.K. *Fiber-Reinforced Composites: Materials, Manufacturing, and design*. CRC Press, 2007.

Mao, K. 2007. A new approach for polymer composite gear design. *Wear* 262(3): 432–441.

Michalski, R.S., Carbonell, J.G., and Mitchell, T.M. *Machine Learning: An Artificial Intelligence Approach*. Springer Science & Business Media, Malaysia, 2013.

Nielsen, L.E., and Landel, R.F. *Mechanical Properties of Polymers and Composites*. Available from: Marcel Dekker Inc., New York, 1994.

Palabiyik, M., and Bahadur, S. 2002. Tribological studies of polyamide 6 and high-density polyethylene blends filled with PTFE and copper oxide and reinforced with short glass fibers. *Wear* 253(3): 369–376.

Pogacnik, A., and Kalin, M. 2012. Parameters influencing the running-in and long-term tribological behavior of polyamide (PA) against polyacetal (POM) and steel. *Wear* 290: 140–148.

Sathees Kumar, S., and Kanagaraj, G. 2016a. Evaluation of mechanical properties and characterization of silicon carbide–reinforced polyamide 6 polymer composites and their engineering applications. *International Journal of Polymer Analysis and Characterization* 21(5): 378–386.

Sathees Kumar, S., and Kanagaraj, G. 2016b. Experimental investigation on tribological behaviours of PA6, PA6-reinforced Al_2O_3 and PA6-reinforced graphite polymer composites. *Bulletin of Materials Science* 39(6): 1467–1481.

Sathees Kumar, S., and Kanagaraj, G. 2016c. Investigation of characterization and mechanical performances of Al_2O_3 and SiC reinforced PA6 hybrid composites. *Journal of Inorganic and Organometallic Polymers and Materials* 26(4): 788–798.

Sathees Kumar, S., and Kanagaraj, G. 2016d. Investigation on mechanical and tribological behaviors of PA6 and graphite-reinforced PA6 polymer composites. *Arabian Journal for Science and Engineering* 41(11): 4347–4357.

Sathees Kumar, S., and Kanagaraj, G. 2017. Effect of graphite and silicon carbide fillers on mechanical properties of PA6 polymer composites. *Journal of Polymer Engineering* 37(6): 547–557.

Serrano, C., Cerrada, M.L., Fernández-García, M., Ressia, J., and Valles, E.M. 2012. Rheological and structural details of biocidal iPP-TiO_2 nanocomposites. *European Polymers* 48: 586–596.

Spiliopoulos, I.K., and Mikroyannidis, J.A. 1998. Soluble phenyl-or alkoxyphenyl-substituted rigid-rod polyamides and polyimides containing m-terphenyls in the main chain. *Macromolecules* 31(4): 1236–1245.

Strong, and Brent, A. *Plastics: Materials and Processing*. Available from: Prentice Hall, New Jersey, 2006.

Sung, N.H., and Suh, N.P. 1979. Effect of fiber orientation on friction and wear of fiber reinforced polymeric composites. *Wear* 53(1): 129–141.

Tiryaki, S., and Aydın, A. 2014. An artificial neural network model for predicting compression strength of heat treated woods and comparison with a multiple linear regression model. *Construction and Building Materials* 62: 102.

Unal, H., and Mimaroglu, A. 2012. Friction and wear performance of polyamide 6 and graphite and wax polyamide 6 composites under dry sliding conditions. *Wear* 289: 132–137.

Young, B.A., Hall, A.L., Pilon, P., Gupta, and Sant, G. 2019. Can the compressive strength of concrete be estimated from knowledge of the mixture pro portions: New insights from statistical analysis and machine learning methods. *Cement Concrete Research* 115: 379.

9 Prediction of Compressive Strength of SCC-Containing Metakaolin and Rice Husk Ash Using Machine Learning Algorithms

Sejal Aggarwal
Manipal University

Garvit Bhargava
National Institute of Technology

Parveen Sihag
Shoolini Universiy

CONTENTS

9.1	Introduction	194
9.2	Machine Learning Approaches	195
	9.2.1 Support Vector Machine (SVM)	195
	9.2.1.1 Multilayer Perceptron (MLP)	195
	9.2.1.2 Gaussian Process Regression (GPR)	195
	9.2.1.3 Random Tree	196
	9.2.1.4 M5P Tree	196
	9.2.1.5 Random Forest (RF)	196
9.3	Methodology and Dataset	196
	9.3.1 Dataset	196
	9.3.2 Model Development	196
	9.3.2.1 Supplied Test Data Method	197
	9.3.2.2 Cross-Validation	198
9.4	Results and Analysis	199
9.5	Discussion	202
9.6	Conclusion	203
References		204

DOI: 10.1201/9781003121954-9

9.1 INTRODUCTION

Concrete, indisputably one of the most marvelous of all human inventions, is used to make houses for the populous to dwell into and construct bridges that span the rivers, dams that control the waters and generate electricity, roads that make trade and communication possible, an endless list of boons to mankind. The recipe for any concrete is cement, water, and aggregates. Concrete hardens as a result of the hydration reaction of cement with water. Concrete has evolved with time to suit the varying needs of mankind. Self-compacting concrete (SCC) is a breakthrough in concrete technology used by large for its distinguished properties from the normal vibrated concrete. Self-compacting concrete gets compacted by its own weight. It has been observed that self-compacting concrete requires relatively lower coarse aggregate and higher cement to maintain its homogeneity as well as its fresh state properties. Admixtures that undergo reactions to form compounds with cement-like properties are therefore widely being tested to be used.

Metakaolin is one such admixture used in the making of self-compacting concrete. The anhydrous calcined form of the clay mineral kaolinite known as metakaolin is observed to improve the engineering properties of concrete. Moghaddam et al. (2015) review that metakaolin increases the compressive strength of concrete mixtures when used in partial replacement. Cement hydration is facilitated by the use of metakaolin as an admixture. The admixture is also observed to decrease creep, drying shrinkage, and chloride permeability. Metakaolin imparting such favorable properties to the concrete mix makes it a widely used admixture). The resulting concrete mix obtained from the partial replacement of metakaolin finds application in multiple walks of life due to the display of superior compressive and flexural strength.

Its properties can also be enhanced by the incorporation of supplementary cementitious materials (SCM) like rice husk ash (RHA). RHA is a viscosity modifying agent that is used to change the properties of concrete, namely, viscosity, cohesiveness, workability, etc. RHA is a by-product of paddy and poses a pollution threat in river water, groundwater, land, and air if not disposed of properly. Using RHA is also a step toward sustainable development by the use of waste material by extension of our technical capabilities. Aboshio et al. (2009) observed that partially replaced RHA increases the compressive strength and the workability of concrete grades 20 and 30. RHA-used concrete also displays increased resistance to chloride penetration. Therefore, the highly pozzolanic RHA is a popular choice to enhance concrete properties.

There is an observable relationship among cement to water ratio, admixture quantities, and aggregate quantities affecting the overall properties of the concrete. One of the most important properties of concrete is its uniaxial compressive strength. A good estimate of the compressive strength of concrete can be obtained after the passage of 28 days of casting it. In this study, the different components incorporated in the formulation of a concrete mix resulting in different compressive strengths are analyzed. The use of soft computing techniques helps in model development to give a mathematical review of the compressive strength of concrete made with different mix ratios. Analytical methods are proven to provide good accuracy within a lesser duration of time and are a cost-effective way to analyze mix ratios. They are not labor exhaustive and are widely being used globally in research work.

The study aims to make use of six soft computing techniques to observe SCC mixes with partially replaced metakaolin and/or RHA using a supplied training data method and a cross-validation method. The study draws attention toward both the methods and their suitability in making predictions of compressive strengths of various SCC mixes.

9.2 MACHINE LEARNING APPROACHES

9.2.1 Support Vector Machine (SVM)

A SVM was introduced by Vapnik (1995). SVM learning algorithms make use of numerical quadratic programming as an inner loop. Linear classifier with the least amount of generalization error is chosen from the infinite number of existing classes. The use of kernel function can be made for nonlinear support vector regression. The method is obtained from statistical learning theory (Platt,1998). Classes are separated and the class having the minimum generalization error is chosen first. In this study, parameters for utilizing SVM-PUK are ω specified as 2, σ as 2, and c as 5 to obtain the optimized results for the data.

9.2.1.1 Multilayer Perceptron (MLP)

McCulloch and Pitts (1943) were the first to have proposed an artificial neural network. MLP consists of three layers, viz., an input layer, a hidden layer, and an output layer. Each layer has a number of neurons present. The number of neurons depends upon the nature of the problem for the input as well as output layers. Error reduction is done by the adjustment of the hidden layer. The performance of each variable on the input layer depends on the value of each neuron. WEKA (3.8) has been used. The training set option is selected with GUI set as true, which generates perceptron with the input layer and output layer, and the hidden layer is kept default. The learning rate is specified as 0.3, the momentum rate is specified as 0.2, and the neural rate is specified as 4 for every hidden unit.

9.2.1.2 Gaussian Process Regression (GPR)

A Gaussian process is a nonparametric flexible method for carrying out nonlinear regression and classification. A Gaussian process belongs to a class of Bayesian nonparametric models. It is a probabilistic method that is kernel-based and provides us to make predictions for noisy and large observational datasets. The most general approach toward forming regression models can be of the following equation type:

$$C(Y(X)) = g(X) \tag{9.1}$$

General regression models are formed on an outcome Y, which is conditional on a set of predictor variables X (Williams & Rasmussen 2006). The covariance of a GP is a matrix and the mean of a GP is a vector. Two kernels GP-PUK and GP-RBF have been utilized. Multiple iterations to optimize the results are done by trial and error. The results for the Pearson VII Kernel function implemented using WEKA 3.8 are received at Gaussian noise specified as 0.04, Ω is specified as 3, and σ is specified as

1. A radial basis kernel function (RBF) was implemented using WEKA 3.8 to obtain results for Gaussian noise specified as 0.045 and spread (gamma) specified as 3.7; the results obtained using RBF had a lower correlation coefficient as compared to PUK, so PUK has been utilized for analysis. The spread parameter helps in defining the width of the bell-shaped curve.

9.2.1.3 Random Tree

Random tree function in WEKA is a supervised decision tree machine learning technique. The random tree starts from a root and the branches split off into various possible solutions. It is a stochastic method where each tree has a uniform sampling possibility. The seed parameter indicates randomization. Arbitrary features k are present at each node. In order to develop precise models, large sets of arbitrary trees are required. Parameters were specified as k equals 10, m as 2, and s as 2 to receive the optimum results.

9.2.1.4 M5P Tree

Introduced by Quinlan et al. (1992), the M5P tree is a technique that combines linear regression functions at the nodes with a decision tree. A decision tree is developed and a linear regression function is assigned at the nodes. M5P is a binary regression model where pre-pruning can be done to deal with splitting at the nodes. The total space of the domain is split into subspaces and a linear regression function is fitted at the nodes of each subspace. In this study, results are obtained with $m = 5$ for an unpruned tree.

9.2.1.5 Random Forest (RF)

RF is a soft computing technique that was introduced by Breiman (2001). It is a collection of decision trees. Observations are selected arbitrarily and each tree is trained by a different subset of data making them each an individual regression function. RF gives final results as an average of all individual trees. This technique safeguards against overfitting. Parameters like the number of trees required in a forest, k (Liaw et al. 2002), and variables chosen at each node for tree development, m, have to be input by the user. To develop the model in this study, k was specified as 2 and m was specified as 0.

9.3 METHODOLOGY AND DATASET

9.3.1 Dataset

A total dataset of 159 mix ratios derived from available literature is subject to two methodologies. The input parameters are cement (kg/m^3), water (kg/m^3), aggregates (kg/m^3), superplasticizer (kg/m^3), metakaolin (kg/m^3), and/or RHA (kg/m^3). The common statistical properties of data are tabulated in Tables 9.1 and 9.2.

9.3.2 Model Development

Models have been developed using the six soft computing approaches stated above. Two ways to develop models are adopted: the supplied test data method and the cross-validation method.

TABLE 9.1
Dataset

Variable	Average	Standard Deviation	Minimum	Maximum
Coarse aggregate (kg/m^3)	794.049	161.106	590.000	1513.000
Fine aggregate (kg/m^3)	828.209	125.482	325.000	1085.000
Water (kg/m^3)	213.161	102.030	121.000	1117.305
Cement (kg/m^3)	423.684	92.492	190.000	681.000
Metakaolin (kg/m^3)	37.962	39.774	0	180.000
Rice husk ash (kg/m^3)	34.711	70.473	0	405.000
Superplasticizer (kg/m^3)	7.401	4.423	0	24.750
Compressive strength (MPa)	46.938	15.471	10.400	107.500

TABLE 9.2
Data sources

Sr. No.	Author	No. of Observations
1	Rukzon and Chindaprasirt (2014)	4
2	Kavitha et al. (2015)	4
3	Chopra et al. (2015)	4
4	Sua-iam and Makul (2013)	5
5	Vivek and Dhinakaran (2017)	5
6	Memon et al. (2011)	6
7	Dinakar and Manu (2014)	6
8	Gholhaki et al. (2018)	6
9	Lenka and Panda (2017)	10
10	Madandoust and Mousavi (2012)	15
11	Gill and Siddique (2017)	16
12	Obunwo et al. (2018)	16
13	Kannan and Ganesan (2014)	17
14	Ghorpade et al. (2015)	18
15	Kannan (2018)	18

9.3.2.1 Supplied Test Data Method

The total dataset (159) is divided arbitrarily into 70% (111), used for training the model, and 30% (48), used for testing the developed model. Performance evaluation of the modeling approaches is done by taking into consideration the correlation coefficients (CC), root mean square error (RMSE), mean absolute error (MAE), relative absolute error (RAE), and root-relative square error (RRSE). The reliability of predictions made by a model is a direct result of the comprehension of the training data. For a successful model, a large variety of data is required. A large variety of data is consequential in establishing the relationship between variables. The input parameters in the model are cement, water, fine aggregates, coarse aggregates,

TABLE 9.3
Features of the Training Dataset

Variable	Average	Standard Deviation	Minimum	Maximum
Coarse aggregate	787.039	153.500	590.000	1513.000
Fine aggregate	831.798	109.653	348.000	1085.000
Water	208.971	81.660	121.000	959.400
Cement	424.794	96.707	190.000	650.000
Metakaolin	37.581	37.258	0	150.000
Rice husk ash	32.721	62.148	0	270.000
Superplasticizer	7.639	4.789	0	24.750
Compressive strength	46.658	14.236	11.900	105.800

TABLE 9.4
Features of the Testing Dataset

Variable	Average	Standard Deviation	Minimum	Maximum
Coarse aggregate	810.259	178.112	590.000	1465.000
Fine aggregate	819.909	157.132	325.000	1071.000
Water	222.853	138.751	148	1117.350
Cement	421.118	82.832	280.000	681.000
Metakaolin	38.845	45.477	0	180.000
Rice husk ash	39.312	87.321	0	405.000
Super plasticizer	6.851	3.410	0	19.100
Compressive strength	47.583	18.158	10.400	107.500

superplasticizers, metakaolin and RHA and the output parameter is compressive strength. The dataset is derived from the literature and out of all the available mixes, certain ambiguous and unsuitable data have been omitted. The features of the training and testing datasets are provided in Tables 9.3 and 9.4, respectively.

The six computing techniques in WEKA version 3.8 were used by running multiple iterations of different user-defined parameters by trial and error. The user-defined parameters, which gave optimum results, are tabulated in Table 9.5. The parameters not specified in the table are used as default.

9.3.2.2 Cross-Validation

Total data are divided into parts by a user-defined number of cross-validation folds. These blocks of data are randomly distributed. In this study, the number of folds was chosen as four. One block of data is selected as a testing dataset and the other three are used for training in the first iteration. In this consideration, another test dataset is chosen from the remaining three and the leftover dataset blocks are used for training. These iterations continue till all four dataset blocks have been used as the testing dataset once while the remaining three in each iteration serve as its training set. This forms four training cases and four testing cases. The whole data are then fed as the

TABLE 9.5
Optimal Values of User-Defined Parameters for Different Machine Learning Algorithms in the Supplied Training Data Method

Model	User Defined Parameters
SVM-PUK	$\omega=2, \sigma=2, c=5$
MLP	Learning rate $=0.3$, Momentum $=0.2$, Neurons $=4$
GPR-PUK	Noise $=0.04$, $\Omega=3$, $\sigma=1$
Random tree	$k=10, m=2, s=2$
M5P tree	Unpruned $=$ **True**, $m=5$
RF	$k=2, m=0, I=50$

TABLE 9.6
Optimal Values of User-Defined Parameters for Different ML Algorithms for the Cross-Validation Method

Model	User Defined Parameter
SVM-PUK	$\omega=1, \sigma=1, c=13$
MLP	Learning rate $=0.05$, Momentum $=0.2$, Neurons $=4$
GPR-PUK	Noise $=0.05$, $\Omega=0.8$, $\sigma=1.2$
Random tree	$k=6, m=1, s=1$
M5P tree	Unpruned $=$ **True**, $m=4$
RF	$k=1, m=7, I=800$

last iteration into the model thus developed. This four-fold cross-validation process offers results for seed value for randomization defined as 1. For the seed value, multiple values of parameters are determined by the trial and error method till optimum results are obtained. The user-defined parameters which gave optimum results are tabulated in Table 9.6.

These parameters are kept unchanged for other randomization seed values. The same procedure is repeated to obtain results for different seed values up to 15, thus giving 15 CC, RMSE, and MAE for 15 seed values. This method delivers more reliable results in the development of models utilizing smaller datasets. Rather than offering a fixed value of results, realistic ranges for the predicted data to lie in between are generated.

9.4 RESULTS AND ANALYSIS

The effectiveness of modeling approaches can be readily assessed by statistical methods. The performance of a model developed is concluded based on its ability to make predictions regarding the strength of SCC. The performance evaluation parameters are representative of how well suited differed predictive algorithms are for the data. The graphs between actual strength and predicted strength of

SCC are plotted in Figure 9.1 for SVM-PUK, MLP, GPR-PUK, random trees, M5P tree, and RF developed for the supplied training method. All values lying on the line of agreement are ideal conditions. The graphs display values lying close to the line of agreement, suggesting that the models developed are satisfactory. By scrutinizing all the graphs, the GPR-PUK model is conclusively having maximum values close to the line of agreement. Table 9.7 indicates the performance evaluation parameters of the six different modeling approaches using the supplied training method. The highest CC for the supplied training method is 0.9315 achieved by GPR-PUK. The GPR-PUK model performs better than other models based on the CC. RF gives the second-best CC of 0.886 suggesting that it is also a well-developed model capable of delivering satisfactory results.

The graphs between actual strength and predicted strength of SCC are plotted in Figure 9.2 for SVM-PUK, MLP, GPR, random tress, M5P tree, and RF developed for the cross-validation method. Values lie in close proximity of the line of agreement

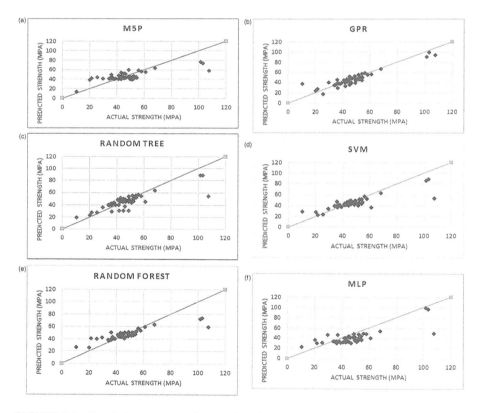

FIGURE 9.1 Graphs between actual strength and predicted strength of SCC for SVM-PUK, MLP, GPR-PUK, random trees, M5P tree, and RF for the supplied training method.

TABLE 9.7
Statistical Measures of Performance Evaluation for the Supplied Training Method

Models	CC	MAE	RMSE	RAE	RRSE
SVM-PUK	0.855	5.705	10.363	0.508	0.629
MLP	0.780	10.092	13.428	0.899	0.746
GPR-PUK	0.932	5.050	6.843	0.450	0.380
Random tree	0.840	6.401	10.209	0.570	0.567
M5P tree	0.732	8.498	12.138	0.757	0.675
RF	0.886	6.474	10.952	0.576	0.609

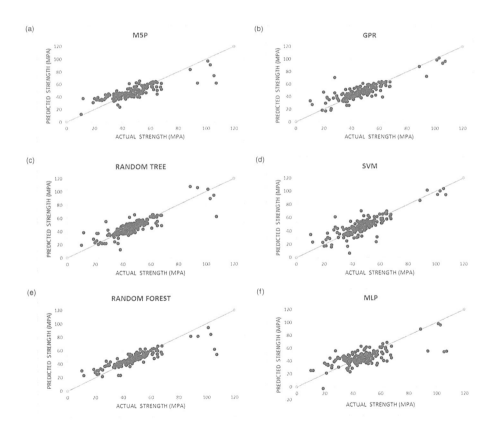

FIGURE 9.2 Graphs between actual strength and predicted strength of SCC for SVM-PUK, MLP, GPR, random tress, M5P tree, and RF.

TABLE 9.8
Statistical Measures of Performance Evaluation for the Cross-Validation Method

Model	Parameter	Mean	Variance	STD. Deviation
SVM-PUK	CC	0.812	0.002	0.045
	MAE	6.035	0.220	0.469
	RMSE	9.010	0.820	0.905
MLP	CC	0.611	0.010	0.102
	MAE	8.564	0.352	0.593
	RMSE	12.331	1.345	1.160
GPR-PUK	CC	0.822	0.002	0.047
	MAE	5.838	0.228	0.477
	RMSE	8.775	0.905	0.951
Random tree	CC	0.714	0.008	0.090
	MAE	6.828	0.529	0.727
	MAE	11.345	3.391	1.841
M5P tree	MAE	0.758	0.002	0.041
	MAE	6.883	0.133	0.365
	MAE	10.121	0.463	0.680
RF	MAE	0.803	0.002	0.040
	MAE	5.621	0.085	0.292
	MAE	9.370	0.570	0.755

suggesting that the models developed are satisfactory. The closer the values lie to the line of agreement, the more reliable the model. The graphs are indicative of well-developed models. Table 9.8 indicates the performance evaluation parameters of the six different modeling approaches using the cross-validation method. In the table, it is observable that the GPR-PUK method yields the highest CC. The CC for GPR is 0.822, which is closely followed by the second highest one of 0.812 yielded by the SVM-PUK model. Figure 9.3 gives graphs on the variation of CC with seed value for the cross-validation method. The seed value corresponding to the best CC for the GPR-PUK method is 2. Seed values for SVM-PUK, MLP, random tree, M5P tree, and RF are 3, 1, 9, 3, and 5, respectively. Figure 9.4 shows graphs on the variation of MAE and RMSE with the change in seed value for the cross-validation method. The optimum CC, MAE, and RMSE obtained by the cross-validation modeling approach can be compared to analyze which of the six soft computing techniques was the most effective. The highest mean CC for cross-validation is 0.8196 achieved by GPR-PUK. Thus, it works the best in comparison to the other techniques.

9.5 DISCUSSION

GPR-PUK emerges out as the most suitable technique for the dataset used. In both the methods using the supplied training dataset as well as the cross-validation dataset,

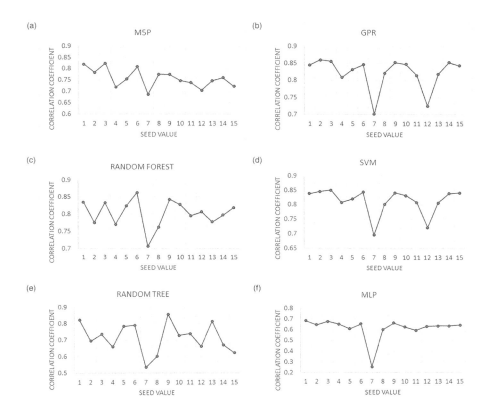

FIGURE 9.3 Graphs on the variation of CC with seed value for the cross-validation method.

GPR-PUK yields the best value of CC. The overall highest value of CC is obtained as 0.932 in the GPR-PUK method using the supplied training dataset. The overall second best CC value is obtained as 0.886 by the RF method using the supplied training dataset. The best value of CC using the cross-validation method is obtained as 0.822 by the GPR-PUK model. The second best CC value in the cross-validation method is obtained as 0.812 by the SVM-PUK model.

9.6 CONCLUSION

This study delves into the effectiveness of soft computation techniques in predicting the compressive strength of SCC containing metakaolin and/or RHA. The CC is highest as 0.932 for GPR-PUK using the supplied training data method. The best mean value for the cross-validation technique is yielded by GPR-PUK as 0.822. Therefore, GPR-PUK portrays the overall best performance among all the soft computing techniques used. The study suggests that GPR-PUK emerges as the best technique.

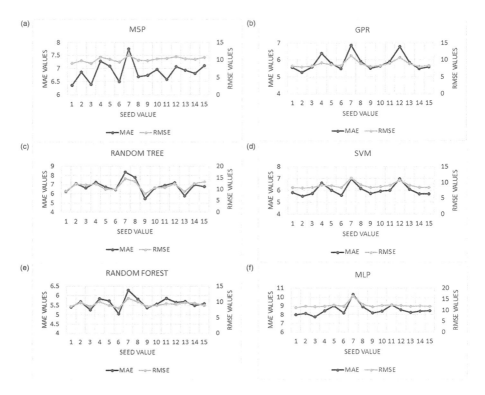

FIGURE 9.4 Graphs on the variation of MAE and RMSE with the change in seed value for the cross-validation method

REFERENCES

A. Aboshio, E. Ogork, D. Balami, Rice husk ash as admixture in concrete, *Journal of Engineering Technology* 4 (2) (2009) 95–101.

L. Breiman, Random forests, *Machine learning* 45 (1) (2001) 5–32.

D. Chopra, R. Siddique, Kunal, Strength, permeability and microstructure of self-compacting concrete containing rice husk ash, Biosystems Engineering 130 (2015) 72–80.

P. Dinakar, S. Manu, Concrete mix design for high strength self-compacting concrete using metakaolin, *Materials & Design* 60 (2014) 661–668.

M. Gholhaki, et al., An investigation on the fresh and hardened properties of self-compacting concrete incorporating magnetic water with various pozzolanic materials, *Construction and Building Materials* 158 (2018) 173–180.

V.G. Ghorpade, K.V. Subash, L.C.A. Kumar, Development of mix proportions for different grades of metakaolin based self-compacting concrete, in: *Advances in Structural Engineering*, Edited by Vasant Matsagar, Springer (2015), pp. 1733–174.

A.S. Gill, R. Siddique, Strength and micro-structural properties of self-compacting concrete containing metakaolin and rice husk ash, *Construction and Building Materials* 157 (2017) 51–64.

V. Kannan, Strength and durability performance of self-compacting concrete containing self-combusted rice husk ash and metakaolin, *Construction and Building Materials* 160 (2018) 169–179.

V. Kannan, K. Ganesan, Chloride and chemical resistance of self-compacting concrete containing rice husk ash and metakaolin, *Construction and Building Materials* 51 (2014) 225–234.

O. Kavitha, V. Shanthi, G.P. Arulraj, P. Sivakumar, Fresh, micro-and macro-level studies of metakaolin blended self-compacting concrete, *Applied Clay Science* 114 (2015) 370–374.

S. Lenka, K. Panda, Effect of metakaolin on the properties of conventional and self-compacting concrete, *Advances in Concrete Construction* 5 (1) (2017) 31–48.

A. Liaw, et al., Classification and regression by random forest, *R News* 2 (3) (2002) 18–22.

R. Madandoust, S.Y. Mousavi, Fresh and hardened properties of self-compacting concrete containing metakaolin, *Construction and Building Materials* 35 (2012) 752–760.

W.S. McCulloch, W. Pitts, A logical calculus of the ideas immanent in nervous activity, *The Bulletin of Mathematical Biophysics* 5 (4) (1943) 115–133.

S.A. Memon, M.A. Shaikh, H. Akbar, Utilization of rice husk ash as viscosity modifying agent in self compacting concrete, *Construction and Building Materials* 25 (2) (2011) 1044–1048.

A.F.K. Moghaddam, R.S. Ravindrarajah, V. Sirivivatnanon, Properties of metakaolin concrete-a review, in: *International Conference on Sustainable Structural Concrete*, 2015, pp. 15–18.

U.E. Obunwo, E.N. Barisua, W. Godfrey, C. Obunwo, Workability and mechanical properties of high-strength self-compacting concrete blended with metakaolin, *SSRG International Journal of Civil Engineering* 5 (10) (2018) 17–22.

J. Platt, Sequential minimal optimization: A fast algorithm for training support vector machines Technical Report MSR-TR-98-14 April 21, 1998 © 1998 John Platt.

J.R. Quinlan, et al., Learning with continuous classes, in: *5th Australian Joint Conference on Artificial Intelligence*, Vol. 92, World Scientific, 1992, pp. 343–348.

S. Rukzon, P. Chindaprasirt, Use of rice husk-bark ash in producing self-compacting concrete, *Advances in Civil Engineering* 2014 (2014) 1–6.

G. Sua-Iam, N. Makul, Utilization of limestone powder to improve the properties of self-compacting concrete incorporating high volumes of untreated rice husk ash as fine aggregate, *Construction and Building Materials* 38 (2013) 455–464.

S. Vivek, G. Dhinakaran, Fresh and mechanical properties of metakaolin-based high-strength scc, *Jordan Journal of Civil Engineering* 11 (2) (2017) 325–333.

V.N. Vapnik, *The Nature of Statistical Learning Theory*. New York: Springer (1995).

C.K. Williams, C.E. Rasmussen, *Gaussian Processes for Machine Learning*, Vol. 2, MIT Press: Cambridge, MA (2006).

10 Predicting Compressive Strength of Concrete Matrix Using Engineered Cementitious Composites: A Comparative Study between ANN and RF Models

Nitisha Sharma and Mohindra Singh Thakur
Shoolini University

Viola Vambol
National Scientific and Research Institute of Industrial Safety and Occupational Safety and Health

Sergij Vambol
Kharkiv Petro Vasylenko National Technical University of Agriculture

CONTENTS

10.1 Introduction 208
10.2 Soft Computing Techniques 209
    10.2.1 Artificial Neural Network (ANN) 209
    10.2.2 Random Forest (RF) 209
10.3 Methodology and Dataset 211
    10.3.1 Dataset 211
    10.3.2 Model Evaluation 211
10.4 Result Analysis 214
    10.4.1 Assessment of the ANN-Based Model 214

10.4.2 Assessment of the RF-Based Model 216
10.4.3 Comparison among the Best-Developed Models 216
10.5 Conclusion 218
References 220

10.1 INTRODUCTION

India is a developing country and shows a boom in industrialization and urbanization. This is one of the reasons for the increase in demand for concrete. Also, since ancient times, concrete plays a significant role in the construction world such as the formation of bridges, roads, or any structural components. Its strength plays an important role because it directly affects the structure. The basic components of concrete are cement (C), fine aggregate (FA), coarse aggregate (CA), and water (w) (Li, 2011). But these constituents are not enough to overcome the drawbacks of concrete. To find a solution, different approaches are performed to find the desired output (Chopra et al., 2018). In a recent scenario, researchers developed engineered cementitious composites to enhance concrete strength properties (Zhang et al., 2017). Different combinations of industrial waste and fibers are used as types of materials, which are found to be useful to get the desired results (Sharma et al., 2020). Various analysts use different combinations of wastes and fibers, which directly affect the strength properties of concrete (Singh et al., 2019).

New components are introduced to overcome the faults in concrete, such as steel fiber is used with polyvinyl alcohol (PVA) fiber in concrete to influence the strength properties of concrete (Zhang et al., 2017). Engineered cementitious composites are used such as polyvinyl alcohol (PVA) fiber in different percentages to get the desired output (Yucel and Kucukarslan, 2016). Soe et al. (2013) used a combination of PVA and steel fiber in the concrete mix to reach the desired output. According to the study, it was concluded that the use of 1.75% and 0.58% of PVA and steel fiber respectively showed better results compared to the composite matrix. Meng et al. (2017) demonstrated that PVA was used with local ingredients to reduce the cost and also to enhance the strength properties of concrete mix.

In the past studies, various soft computing techniques were applied to input parameters of concrete mix to predict the desired output. Strength properties of concrete were analyzed for a different number of days by a linear mathematical model. Ahmed (2012) used cement, time, superplasticizer (SP), etc. as input parameters to reach a conclusion. It was found that the predicted model was able to estimate the desired output. Topcu and Saridemir (2008) analyzed the data from past studies where input parameters were taken as fly ash with a primary constituent of concrete mix and fuzzy logic (FL), and neural network soft computing techniques were applied to predict the output (strength property). In the study, it was found that a neural network predicts better results to reach the desired outcome compared to FL.

Based on the published manuscripts, the objective of this chapter is to compare and find the better-performed model, with variation in input parameters. Soft computing techniques, such as artificial neural network (ANN) and random forest (RF) were applied on ten different models to reach the goal of the study. Cement (C), fine aggregate (FA), water, superplasticizers/high range water reducers (SP/HRWR), fly ash, polyvinyl alcohol (PVA), steel fiber, and curing days were used as input variables to achieve the output of compressive strength (CS) for concrete.

10.2 SOFT COMPUTING TECHNIQUES

Various soft computing techniques have been adopted in recent studies to solve the complexity in engineering design. Soft computing techniques, such as ANN, RF, random tree, FL, and M5P, were applied in different engineering problems to reach the desired goal (Thakur et al., 2021). The construction industry is dependent on the properties of the materials used, so to reach the goal, operational parameters, for example, cement, sand, aggregates, water, additives (fibers, waste, etc.) play a crucial role. To estimate the properties of the concrete matrix, different models were developed (Chopra et al., 2018; Sepahvand et al., 2019; Han et al., 2019; Goldberg, 2017; Nhu et al., 2020). These techniques are able to find a solution for engineering problems, for example, Topcu and Saridemir (2008) used the ANN and FL techniques for the estimation of the strength properties of concrete and found that ANN gave better results compared to the FL technique.

In this study, ANN and RF techniques were applied to understand the relationship of operational parameters, i.e., cement, fine aggregates, water, SP/HRWR, fly ash, PVA, steel fibers, and curing days, to achieve the target strength of the concrete.

10.2.1 Artificial Neural Network (ANN)

An ANN is a soft computing technique that simulates and processes information. This artificial intelligence technique processes complex information to reach the desired goal. ANN processes the input parameters, which behave like neurons. These neurons are interconnected and are able to find out the best-estimated output. There are three types of layers, namely, input layer (include all input variables), hidden layer (number of neurons), and output layer (target). A number of neurons are present in hidden layers; a complex structure is formed when these neurons are connected with the input layer. The total dataset splits into two subsets: one is the training subset and the other one is the testing subset. The training dataset contains 70% of the data from the total dataset, which includes all the maximum and minimum values of all the input and output parameters, the remaining 30% is used as the testing subset. This technique is applied on Weka 3.8.4 software to predict the results. Figure 10.1 shows the ANN-based model, which includes three layers. The first layer contains all the input parameters, the second layer is the hidden layer with six neurons, and the third layer is the output layer. The combination of these entire layers forms an ANN model. The best model is selected by the trial and error method with parameters which are user-defined such as learning rate = 0.2, momentum = 0.1, no. of iteration = 7000, and hidden layer = 1 with no. of neurons = 6.

10.2.2 Random Forest (RF)

RF is an artificial intelligence technique, which was first introduced by Breiman (2001) to find a solution for complex problems by classification and regression. The RF model

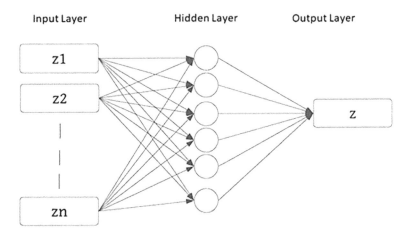

FIGURE 10.1 Structure of an artificial neural network.

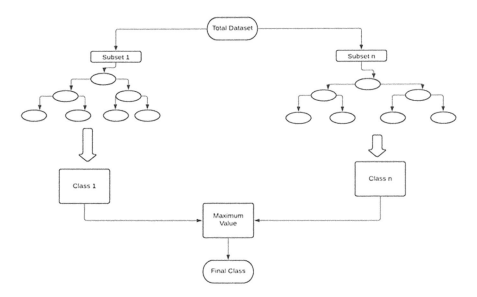

FIGURE 10.2 Structure of random forest.

contains a number of trees as shown in Figure 10.2, which grow together, and a forest was formed where data were randomly assigned to each tree (Thakur et al., 2021). Data were divided into two parts, i.e., 70% and 30% for training and testing subsets respectively from the total dataset. The random distribution of data to each tree is known as the bagging process (Chopra et al., 2018). For the regression process, input boundaries were utilized at every node for the development of a tree and selected variables had been approached for the best split on every node (Sepahvand et al., 2019). Upon random replacement of each variable, the importance of input parameters was determined.

The quality of the model can be evaluated by doing some alterations in the error for the testing dataset. Weka 3.8.4 software was used for the RF; the model was prepared with the number of iterations = 1000, no. of features (k) = 0, and seed (s) = 2.

10.3 METHODOLOGY AND DATASET

10.3.1 DATASET

Collecting data and selecting input and output variables are important parts of the prediction study. Previous publications were used to collect the data. A total of 126 observations were collected from the literature and the details of the observations are listed in Table 10.1. Out of 126, 88 were used for training and the remaining 38 were used for model testing/validation. Features of the dataset for training and testing are listed in Table 10.2. For model development, different input combination-based models were prepared. Cement (C), fine aggregate (FA), water (w), SP/HRWR, fly ash, PVA, steel fiber, and curing days were adopted as input parameters to achieve the output where compressive strength (CS) was taken as an output parameter. Table 10.3 shows the details of different combinations of input parameters. A flow chart is also plotted in Figure 10.3 for a better understanding of the methodology.

Different models are developed with variations in input combinations to find the best-suited model for the estimation of the compressive strength of the concrete.

10.3.2 MODEL EVALUATION

After a combination of various variable sub-sets for making ten models, evaluation of parameters has to be carried out for checking the accuracy. Evaluating parameters used in this study are correlation coefficient (CC), mean absolute error (MAE), root mean square error (RMSE), relative absolute error (RAE), and root relative squared error (RRSE).

1. $$CC = \frac{s\left(\sum_{i=1}^{s} SR\right) - \left(\sum_{i=1}^{s} S\right)\left(\sum_{i=1}^{s} R\right)}{\sqrt{\left[s\sum_{i=1}^{s} S^2 - \left(\sum_{i=1}^{s} S\right)^2\right]\left[s\sum_{i=1}^{s} R^2 - \left(\sum_{i=1}^{s} R\right)^2\right]}} \quad (10.1)$$

2. $$RMSE = \sqrt{\frac{1}{s}\left(\sum_{i=1}^{s}(R-S)^2\right)} \quad (10.2)$$

3. $$RAE = \frac{\sum_{i=1}^{s}|S-R|}{\sum_{i=1}^{s}\left(|S-\bar{S}|\right)} \quad (10.3)$$

TABLE 10.1
Detail of a Dataset

S. No	1	2	3	4	5	6	7	8	Total
Author's name	Soe et al.	Meng et al.	Al-Gemeel et al.	Yucel and Kucukarslan	Zhang et al.	Mohammed et al.	Khed et al.	Mohammed et al.	
Year	2013	2017	2018	2016	2016	2018	2020	2019	
Total dataset	4	5	4	10	20	13	50	20	126

TABLE 10.2
Features of a Dataset

Statistics	Cement	Fine Aggregate	Water	SP/HRWR	Fly Ash	PVA%	Steel %	Curing Days	Compressive Strength (MPa)
					Training Dataset				
Minimum	1.0000	0.0000	0.1800	0.0060	0.0000	0.0000	0.0000	7.0000	39.3000
Maximum	1.0000	2.6560	0.5800	3.1900	5.2440	2.4100	2.6200	28.0000	105.5000
Mean	1.0000	0.7042	0.3363	0.5829	1.1685	1.2006	0.3586	24.4205	68.4764
Standard deviation	0.0000	0.4218	0.1184	0.6584	0.7891	0.6922	0.5019	7.9419	14.6544
					Testing Dataset				
Minimum	1.0000	0.0000	0.1800	0.0090	0.0000	0.0000	0.0000	7.0000	40.0000
Maximum	1.0000	1.2330	0.5800	2.9100	2.4290	2.2000	2.0000	28.0000	102.8000
Mean	1.0000	0.7237	0.3293	1.0737	1.2433	1.0621	0.1803	26.8947	78.4892
Standard deviation	0.0000	0.3259	0.0805	0.8882	0.4868	0.7473	0.3877	4.7522	15.3207

TABLE 10.3
Combination of Various Variable Sub-Sets

Sr. No.	Input Combinations	Output	Model No.
1	C+FA+W+SP/HRWR+ Fly ash+ PVA%+ Steel%+Curing days	Compressive strength	M1
2	C+FA+W+ Fly ash+ PVA%+Steel%+Curing days	Compressive strength	M2
3	C+FA+W+SP/HRWR+ PVA%+ Steel%+Curing days	Compressive strength	M3
4	C+FA+ SP/HRWR+ Fly ash+ PVA%+ Steel%+Curing days	Compressive strength	M4
5	C+ W+SP/HRWR+ Fly ash+ PVA%+ Steel%+Curing days	Compressive strength	M5
6	C+FA+W+SP/HRWR+ Fly ash+ Steel%+ Curing days	Compressive strength	M6
7	C+FA+W+SP/HRWR+ Fly ash+ PVA%+ Curing days	Compressive strength	M7
8	C+FA+W+ PVA%+Steel%+Curing days	Compressive strength	M8
9	C+FA+W+SP/HRWR+ Fly ash+ Curing days	Compressive strength	M9
10	C+FA+W+ Steel%+Curing days	Compressive strength	M10

4. $\text{RRSE} = \sqrt{\dfrac{\sum_{i=1}^{s}(S-R)^2}{\sum_{i=1}^{s}(|S-\bar{R}|)^2}}$ (10.4)

5. $\text{MAE} = \dfrac{1}{s}\left(\sum_{i=1}^{s}|R-S|\right)$ (10.5)

S = Observed values
$\bar{S}$ = Average of observed value
R = Predicted values
s = Number of observations

CC is one of the significant evaluating parameters that describe the performance of the models. It ranges from −1 to +1 beyond these values and shows the error in model relations. The higher the value of CC, the better the results. Similarly, RMSE and MAE are helpful in evaluating errors. The minimum the value of errors, the better the predicted results (Nhu et al., 2020). The percentage error is RAE and RRSE is equally significant. If the relative error is high, the calculated value is very small, which means that the prediction is not good for the desired output and vice versa.

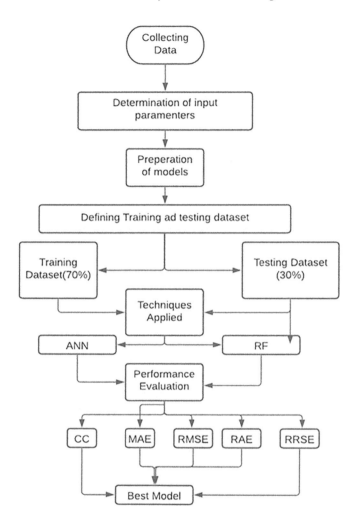

FIGURE 10.3 Flow chart of methodology.

10.4 RESULT ANALYSIS

10.4.1 Assessment of the ANN-Based Model

To check the performance of different input variable combinations, five statistic performance evaluation measures are applied, which include CC, MAE, RMSE, RAE, and RRSE. These measures were calculated for both the training and testing datasets. Results of these performance measure values are listed in Table 10.4 for both stages. Results of Table 10.4 suggest that M-7 is performing better than the other input combination-based models with CC being 0.8731, MAE being 8.2847, RMSE being 9.9163, RAE being 54.31%, and RRSE being 54.69%. A scatter plot is drawn (Figure 10.4) for the comparison among actual and predicted values using

TABLE 10.4
Statistic Quantitative Results of ANN

Model No.	Training					Testing				
	CC	MAE (MPa)	RMSE (MPa)	RAE (%)	RRSE (%)	CC	MAE	RMSE	RAE (%)	RRSE (%)
M1	0.9611	3.4927	4.5138	30.0143	30.9786	0.8734	12.3666	15.3454	81.0635	84.6200
M2	0.9015	5.4718	7.5919	47.0217	52.1031	0.7887	13.8635	16.0828	90.8757	88.6935
M3	0.9218	4.6531	6.0381	39.9858	41.4392	0.7469	20.8963	27.0117	136.9757	148.9643
M4	0.9614	2.9802	4.1484	25.6101	28.4703	0.6043	18.0989	22.1968	118.6387	122.4110
M5	0.9006	5.3874	6.7589	46.2957	46.3864	0.6145	13.6734	21.0948	89.6293	116.3340
M6	0.9532	4.3561	5.6227	37.4339	38.5886	0.8084	8.0607	10.2113	52.8382	56.3136
M7	0.9254	5.8509	7.4989	50.28	51.47	0.8731	8.2847	9.9163	54.31	54.69
M8	0.9089	4.3510	6.3650	37.3895	43.6829	0.6410	13.2239	15.2613	86.6833	84.1631
M9	0.9365	6.2600	7.5079	53.7943	51.5269	0.7943	9.4939	11.8890	62.2327	65.5654
M10	0.8457	5.9726	8.4084	51.3248	57.7069	0.7909	11.7339	14.5813	76.9157	80.4129

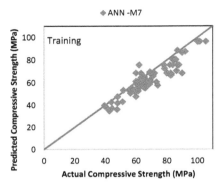
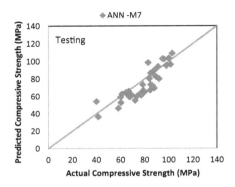

FIGURE 10.4 Scatter graphs of observed and predicted compressive strengths of the best model in the training and testing datasets by the ANN technique.

the M-7 ANN model. Figure 10.4 and Table 10.4 suggest that the M-7 ANN model is suitable for the prediction of the compressive strength of concrete.

10.4.2 Assessment of the RF-Based Model

The performance of different input variable combinations is evaluated by applying five statistic measures, which include CC, MAE, RMSE, RAE, and RRSE. These measures were calculated for both the training and testing datasets. Results of these performance measure values are listed in Table 10.5 for both stages. Results of Table 10.5 suggest that M-4 is performing better than other input combination-based models with CC being 0.9049, MAE being 6.1552, RMSE being 7.9275, RAE being 40.3472%, and RRSE being 43.7186%. A scatter plot is drawn (Figure 10.5) for the comparison among actual and predicted values using the M-4 ANN model. Figure 10.5 and Table 10.5 suggest that the M-4 ANN model is suitable for the prediction of the compressive strength of the concrete matrix.

10.4.3 Comparison among the Best-Developed Models

In this chapter, soft computing techniques were applied to estimate the compressive strength of concrete where basic constituents of concrete were mixed with fly ash, PVA, and steel fibers. Performance was also evaluated for training and testing subsets by applying statistics measures and a comparison of the results is listed in Table 10.6. A scatter plot was also prepared as shown in Figure 10.6, for comparing the actual and predicted values of the compressive strength for the best models obtained from ANN and RF techniques. Figure 10.6 shows that relative error was minimum in the RF-M4 model for both training and testing datasets. It shows that the bandwidth of the error is close to the zero error line in the RF-M4 model compared to ANN-M7 and predicted RF-M4 values are closer to the actual

TABLE 10.5
Statistic Quantitative Results of RF

Model No.	Training					Testing				
	CC	MAE	RMSE	RAE (%)	RRSE (%)	CC	MAE	RMSE	RAE (%)	RRSE (%)
M1	0.9868	1.9915	2.6607	17.1138	18.2606	0.8975	6.4217	8.4124	42.0945	46.3926
M2	0.9858	2.0585	2.7909	17.6895	19.1537	0.7508	10.4928	13.111	68.7806	72.3048
M3	0.986	2.1016	2.805	18.0603	19.2508	0.8617	7.2509	9.0515	47.5299	49.9174
M4	0.9866	1.9962	2.7336	17.1545	18.7605	0.9049	6.1552	7.9275	40.3472	43.7186
M5	0.9807	2.7631	3.4004	23.7446	23.3372	0.8963	7.003	8.78	45.9045	48.4198
M6	0.9768	2.2179	3.2973	19.0590	22.6297	0.8658	6.2011	8.4459	40.6484	46.5776
M7	0.9665	2.5998	3.8529	22.3413	26.4427	0.8763	6.6004	8.5811	43.2655	47.3234
M8	0.9662	275.65%	395.53%	23.6878	27.1455	0.5726	12.8233	15.093	84.0569	83.2352
M9	0.9588	278.24%	418.21%	23.9101	28.7022	0.8602	6.3579	8.3837	4167.6000	46.2347
M10	0.955	302.60%	441.90%	26.0032	30.3277	0.4499	11.8953	15.6709	77.9738	86.4220

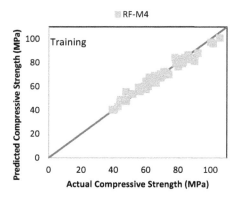
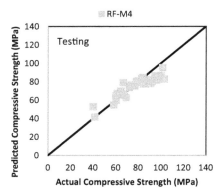

FIGURE 10.5 Scatter graphs of observed and predicted compressive strengths of the best model in the training and testing datasets by the RF technique.

TABLE 10.6
Comparison of Model Prediction

Models	CC	MAE (MPa)	RMSE (MPa)	RAE (%)	RRSE (%)	Order
RF-M4	0.9049	6.1552	7.9275	40.3472	43.7186	1
ANN-M7	0.8731	8.2847	9.9163	54.31	54.69	2

values. Also, according to the statistical analysis, RF-M4 predicts better results among all the models as listed in Table 10.6. Results of the CC showed that both the techniques are acceptable because the CC value is 0.9049 and 0.8731 for RF and ANN, which means fewer errors for estimating the compressive strength of concrete. RF-M4 shows better results with the highest value of CC (0.9049) followed by ANN-M7 with CC (0.8731), also RMSE results show RF-M4 has the best values for accurate data with the minimum error value of 7.9275 followed by ANN-M7 (9.9163). Similarly, RRSE has a minimum error value for RF-M4 (43.7186%) followed by ANN-M7 (54.69%).

10.5 CONCLUSION

In this chapter, a number of models were designed by applying variation in the input parameters and soft computing techniques were applied such as ANN and RF for the estimation of compressive strength of the concrete. For the evaluation of results, performance measures such as CC, RMSE, MAE, RAE, and RRSE were checked for all designed models. Results of this study are summarized below:

1. RF-M4 was found to be the most suitable for the estimation of the compressive strength of concrete, where cement, fine aggregate, SP/HRWR, fly ash, PVA%, steel fiber %, and curing days were taken as input parameters.

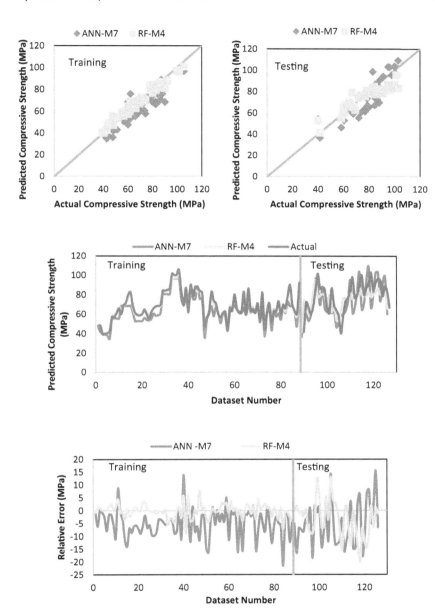

FIGURE 10.6 Comparison between ANN-M7 and RF-M4, predicted and actual values.

2. Results showed that best performance was given by the RF-M4 model with highest CC value = 0.9049 and lower RMSE value = 7.9275 MPa followed by the ANN-M7 model with CC value = 0.8731 and RMSE value = 9.9163 MPa.
3. The scatter diagram shows that the RF-M4 model is a good fit for the observed data and thus for the estimation of compressive strength of concrete mix due to low error bandwidth and close to the zero error line.

REFERENCES

Ahmed M.S. Statistical modelling and prediction of compressive strength of concrete. *Concrete Research Letters.* 2012; 3(2). doi: 10.6084/M9.FIGSHARE.105905.

Al-Gemeel A.N., Zhuge Y., and Youssf O. Use of hollow glass microspheres and hybrid fibres to improve the mechanical properties of engineered cementitious composite. *Construction and Building Materials.* 2018; 171(2018): 858–870. https://doi.org/10.1016/j.conbuildma t.2018.03.172.

Breiman L. Random forests. *Machine Learning.* 2001; 45(0): 5–32.

Chopra P., Sharma R.K., Kumar M., and Chopra T. Comparison of machine learning techniques for the prediction of compressive strength of concrete. *Advances in Civil Engineering.* 2018. doi: 10.1155/2018/5481705.2

Goldberg Y. Neural network methods for natural language processing. *Synthesis Lectures on Human Language Technologies.* 2017; 10(1): 1–309. doi: 10.2200/S00762ED1V01Y201703HLT037.

Han Q., Gui C., Xu J., and Lacidogna G. A generalized method to predict the compressive strength of high-performance concrete by improved random forest algorithm. *Construction and Building Materials.* 2019; 226(0): 734–742. doi: 10.1016/j.conbuildmat.2019.07.315.

Khed V.C., Mohammed B.S., Liew M.S., and Zawawi N.A.W.A. Development of response surface models for self-compacting hybrid fibre reinforced rubberized cementitious composite. *Construction and Building Materials.* 2020; 232(2020): 117191. https://doi.org/10.1016/j.conbuildmat.2019.11719.

Li Z., *Advanced Concrete Technology.* John Wiley & Sons, 2011.

Meng D., Huang T., Zhang Y.X., and Lee C.K. Mechanical behavior of a polyvinyl alcohol fibre reinforced engineered cementitious composite (PVA-ECC) using local ingredients. *Construction and Building Materials.* 2017; 141(2017): 259–270. http://dx.doi.org/10.1016/j.conbuildmat.2017.02.158

Mohammed B.S., Achara B.E., Liew M.S., Alaloul W.S., and Khed V.C. Effects of elevated temperature on the tensile properties of NS-modified self-consolidating engineered cementitious composites and property optimization using response surface methodology (RSM). *Construction and Building Materials.* 2019; 206(2019): 449–469. https://doi.org/10.1016/j.conbuildmat.2019.02.033.

Mohammed B.S., Khed V.C., and Liew M.S. Optimization of hybrid fibres in engineered cementitious composites. *Construction and Building Materials.* 2018; 190(2018): 24–37. https://doi.org/10.1016/j.conbuildmat.2018.08.188.

Nhu V.H., Shahabi H., Nohani E., Shirzadi A., Ansari N.A., Bahrami S., Miraki S., Geertsema M., and Nguyen H. Daily water level prediction of zrebar lake (Iran): A comparison between M5P, Random Forest, Random Tree and Reduced Error Pruning Trees Algorithms. *International Journal of Geo-Information.* 2020; 9: 479. doi: 10.3390/ijgi9080479.

Sepahvand A., Singh B., Sihag P., Samani A.N., Ahmadi H., and Nia S.F. Assessment of the various soft computing techniques to predict sodium absorption ratio (SAR). *ISH Journal of Hydraulic Engineering.* 2019. doi: 10.1080/09715010.2019.1595185.

Sharma N., Thakur M.S., Goel P.L., and Sihag P. A review: Sustainable compressive strength properties of concrete mix with replacement by marble powder. *Journal of Achievements in Materials and Manufacturing Engineering.* 2020; 98(1): 11–23. doi: 10.5604/01.3001.0014.0813.

Singh B., Sihag P., Tomar A., and Sehgad A. Estimation of compressive strength of high-strength concrete by random forest and M5P model tree approaches. *Journal of Materials and Engineering Structures.* 2019; 6(2019): 583–592.

Soe K.T., Zhang Y.X., Zhang L.C. Material properties of a new hybrid fibre-reinforced engineered cementitious composite. *Construction and Building Materials.* 2013; 43(2013): 399–407. doi: 10.1016/j.conbuildmat.2013.02.021.

Thakur M.S., Pandhiani S.M., Kashyap V., Upadhya A. and Sihag P. Predicting bond strength of FRP bars in concrete using soft computing techniques. *Arabian Journal of Science and Engineering.* 2021. https://doi.org/10.1007/s13369-020-05314-8.

Topcu I.K., and Saridemir M. Prediction of mechanical properties of recycled aggregate concretes containing silica fume using artificial neural networks and fuzzy logic. *Computational Materials Science.* 2008; 42(2008): 74–82. doi: /10.1016/j.commatsci.2007.06.011.

Yucel H.E., and Kuçukarslan O. Effect of PVA fiber ratio on mechanical and dimensional stability properties of engineered cementitious composites. *Journal of International Scientific Publications.* 2016; 10(0): 595–601. https://citeseerx.ist.psu.edu/viewdoc/download?doi=10.1.1.1065.7025&rep=rep1&type=pdf

Zhang J., Wang Q., and Wang Z. Properties of polyvinyl alcohol-steel hybrid fiber-reinforced composite with high-strength cement matrix. *Journal of Materials in Civil Engineering.* 2017; 29(7). https://doi.org/10.1061/(ASCE)MT.1943-5533.0001868

11 Estimation of Marshall Stability of Asphalt Concrete Mix Using Neural Network and M5P Tree

Ankita Upadhya and Mohindra Singh Thakur
Shoolini University

Siraj Muhammed Pandhian
University College Jubail

Shubham Tayal
SR University

CONTENTS

11.1 Introduction 223
11.2 Machine Learning Techniques 225
    11.2.1 Artificial Neural Network..225
    11.2.2 M5P Model 226
11.3 Methodology and Dataset..227
    11.3.1 Dataset Collection..227
    11.3.2 Performance Assessment of Parameters..227
11.4 Result Analysis 229
    11.4.1 Assessment of the ANN-Based Model 229
    11.4.2 Assessment of the M5P Based Model 229
11.5 Comparison between Best Developed Models 230
11.6 Conclusion 234
References..234

11.1 INTRODUCTION

Asphalt concrete pavement mixes involve the selection and proportioning of materials to obtain a desired mix that will produce a robust pavement with the ability to carry the predicted traffic loads without undergoing excessive distortion or displacement (Saif et al., 2013). An increase in rutting depth accelerates the damage due to the presence of moisture in asphalt pavement and reduces the drainage capacity, which also leads to fatigue cracking. The moisture introduces rutting, which includes three main pavement failures such as thermal cracking, fatigue, and rutting, which is the most damaging failure mechanism in asphalt pavements (Azarhoosh and Koohmishi, 2021). To deal with these problems, many technologies and additives are being presented to enhance the crack resistance of asphalt; among those, fiber reinforcement is an essential measure to improve the asphalt pavement performance (Ziari and Moniri, 2019). Mostly in worldwide some polymers are added in asphalt for increasing its durability, which influences the binder properties positively by increasing the elasticity, cohesion, and stiffness of the pavement (Sohel et al., 2020). In asphalt mixtures, there are several methods for calculating the resistance against the cracking due to low temperature, using rupture mechanics which is one of the most consistent practices used in the pavement design (Hassan et al., 2020). Since the last decade, many engineers have assessed the phenomenon of rutting in asphalt mix pavement. Researchers, in previous studies, have proposed many models to predict rutting potential with a mechanistic-empirical approach with many limitations (Azarhoosh and Koohmishi, 2021).

For each specific pavement, it is not easy to attain reliable rut resistance estimates for each specific section for various parameters in the mix design due to tediousness and high-cost testing in the laboratory (Majidifard et al., 2021). It has been found that the ANNs are capable of simplifying the complexities in relationships between input data and output dataset and also satisfy the model prediction for defining the stiffness modulus, of asphalt concrete mix. After the completion of the experimental mix design process, the formulation of the asphalt mix most suited is recognized, so as to satisfy the utility of the design pavement (Nicola et al., 2018).

Nowadays, artificial intelligence (AI) is widely used to find out the problems related to pavement deterioration. Machine learning algorithms are widely used to develop models and overcome data challenges to give accurate and robust vulnerability prediction of models (Nhu et al., 2020). These artificial intelligence techniques neglect physical relationships in the input and the output variables. This implies that the input parameters, which are used to predict the data should be in between the range of trained data (Bui et al., 2018). The learning ability of the ANN model brings together the fuzzy system linguistic representation (Morova et al., 2017). Distortion and stability values are identified from these tests, and the outcomes from the test inputs are taken for the development of the model (Dao et al., 2020).

Moreover, artificial neural network (ANN) models are easier to execute and they are available with various statistically programmed software (Qadir et al., 2020). For the development of the model, 13 experimental results were used and compared with reference to correlation and error to determine the performance of the best-suited model and a good correlation of the input parameters was found (Serin et al., 2013). It has been observed that ANN's analysis tools are for the fast and accurate prediction of the deflection and under critical conditions of the flexible pavements subjected to typical highway loadings (Saffarzadeh and Heidaripanah, 2009). The performance assessment of developed models was calculated and related to ANN models. For the prediction of dynamic modulus, ANN models are widely used, and it has been observed that the models developed by using the M5P algorithm-based model have outperformed in comparison to previously developed models (Behnood and Daneshvar, 2020). It has been found that the prediction of input parameters observed by using the ANN model shows a better relationship between the experimental data and the performance evaluation of the 4-14-3 architecture was more appropriate than other architectures (Reddy, 2018). An ANN tool is successfully applied to predict the field permeability of hot mix asphalt pavement course and it shows that the ANN-based model has better performance in assessing the accurate prediction of field permeability of hot mix asphalt (Ozgan, 2011). Keeping in view, the improved performances of M5P and ANN techniques are used in various fields to check the statistical performance of models. In this, ANN and M5P are assessed for predicting the Marshall stability of asphalt concrete mix with different input combination-based models to find out the best input parameter with the best-suited model (Mohammed et al., 2020).

11.2 MACHINE LEARNING TECHNIQUES

Various techniques related to machine learning have been widely used to resolve complex problems in the construction fields. In this study, ANN, and M5P techniques were used to analyze the best-suited correlation between the input parameters with the different combinations used, i.e., the type of bitumen, bitumen content (BC), voids, voids in mineral aggregates (VMA), and voids filled with asphalt (VFA) to attain the output parameter, i.e., Marshall stability of asphalt concrete mix.

11.2.1 Artificial Neural Network

A neural network is a machine learning technique that is based on the assembly of algorithms and works on neurons. In an ANN, many multilinear algorithms perform together progression and articulate multiplex input variables (Nguyen et al., 2020). ANN has one or more hidden layers, i.e., the number of inputs and one output layer, each layer containing several nodes and the weighted connection between these layers signifies the connection among the nodes (Sihag et al., 2018). The ANN technique is applied on the training dataset and testing dataset and the validation is done on both the divided dataset. This technique helps in solving complex problems rapidly and calculating the results accurately (Saffarzadeh and Heidaripanah, 2009). However, generating a realistic ANN network requires empirical experiences, and its effectiveness is entirely based on the trial-and-error process. This approach is blackbox and therefore, the nature of the prediction equation is not known. In this study, the performance of the ANN model has been done to obtain desired output by taking five input parameters (Vyas et al., 2020). To estimate the statistical performance of the dataset, Weka 3.8.4 software is used to analyze the performance of the dataset. The basis of the selection of various models performing with a number of hidden layers parameters is presented in Figure 11.1.

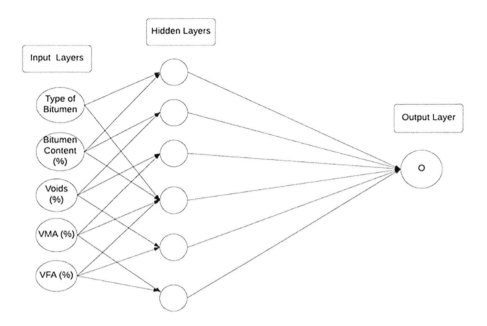

FIGURE 11.1 ANN structure with different input parameters.

11.2.2 M5P Model

M5P trees are essentially dependent on the regression tree model, which can deliver perpetual mathematical credits by utilizing linear regression (Nouri et al., 2020). At a terminal leaf node, there is a binary decision tree that has tendency to assess the linear equations with continuous mathematical features (Thakur et al., 2021). The binary tree assigns linear regression features to the terminal leaf node and finds a way to fit in a model of linear regression on each sublocation by separating the dataset into different areas (Mohammed et al., 2020). The changeability is calculated by the values of standard deviation reaching the root node through the branch, which evaluates the estimated reduction error in testing each node variable. The preference is given to the variable responsible for maximizing reduction in expected error. The splitting process will stop if the variation is encountered in instances that are reaching the node. The reduction in the standard deviation of each terminal node is determined by the following equation (Sihag et al., 2020a):

$$\text{SDR} = \text{sd}(Q) - \sum_{i}^{n}\left(\frac{|Q_i|}{|Q|}\text{sd}(Q_i)\right) \quad (11.1)$$

where Q is the set of examples arriving at the node, Q_i depicts the ith output of a subset of Q, and sd is the standard deviation. Weka 3.8.4 software is used in this study for the M5P modeling.

11.3 METHODOLOGY AND DATASET

11.3.1 Dataset Collection

The dataset is collected from Nicola et al. (2018). The total dataset incorporating 90 readings was statically analyzed and divided randomly into two subsets, i.e., training dataset and testing dataset. Out of these, a total of 63 observations are taken for the training dataset and the rest of the observations (23) is taken for the testing dataset, which is used for the validation of the model development. The features of the training and testing dataset are shown in Table 11.1. For the validation of the model, different input combinations of models were prepared which contain the type of bitumen, BC, voids (%), VMA, VFA, and for the prediction of Marshall stability (KN), which is an output parameter. The details of these different input combination-based models are listed in Table 11.2. A flow chart is graphically represented in Figure 11.2 for understanding the methodology precisely.

All these combinations are applied to construct a model for the prediction of Marshall stability as an output.

11.3.2 Performance Assessment of Parameters

Five statistical indices parameters for performance assessment of various models for estimating the output, such as correlation coefficient (CC), root mean square error (RMSE), relative-absolute error (RAE), Root relative squared error (RRSE), and mean-absolute error (MAE), are calculated to determine the efficacy of machine learning techniques. The range of CC is from −1 to +1. However, zero CC indicates

TABLE 11.1
Features of a Dataset

Statistics	Bitumen Content (%)	Voids (%)	VMA (%)	VFA (%)	Marshall Stability (KN)
			Training Dataset		
Minimum	3.8000	0.4000	12.1000	48.0000	7.8000
Maximum	5.8000	9.7000	20.4000	97.4000	16.5000
Mean	4.7871	5.2698	15.8730	67.5937	12.2254
Standard deviation	0.4895	2.2785	2.3036	11.8019	1.9455
Kurtosis	0.0225	−0.6470	−0.9282	−0.0376	−0.5951
Skewness	−0.1568	−0.1174	0.5440	0.5169	−0.0114
			Testing Dataset		
Minimum	3.8000	0.5000	13.1000	48.0000	8.2000
Maximum	5.8000	9.0000	20.2000	96.3000	15.5000
Mean	4.6637	5.9963	16.3111	64.1963	12.8556
Standard deviation	0.5216	2.2930	2.1387	12.1961	1.7926
Kurtosis	−0.2478	0.0746	−1.0937	0.8557	0.7762
Skewness	0.1672	−0.7244	0.4450	1.0265	−0.8134

TABLE 11.2
Combination of Various Variable Subsets

Model No.	Type of Bitumen	Bitumen Content (BC)	Voids (%)	VMA (%)	VFA (%)	Output (KN)
M1	✓	✗	✓	✓	✓	Marshall stability
M2	✓	✓	✓	✓	✓	Marshall stability
M3	✗	✓	✓	✓	✓	Marshall stability
M4	✓	✓	✗	✓	✓	Marshall stability
M5	✓	✓	✓	✗	✓	Marshall stability
M6	✓	✓	✓	✓	✓	Marshall stability
M7	✓	✗	✗	✓	✓	Marshall stability
M8	✓	✓	✗	✗	✓	Marshall stability
M9	✓	✓	✓	✗	✗	Marshall stability
M10	✗	✓	✓	✓	✗	Marshall stability

no relationship between the actual and predicted data (Sihag et al., 2020b). Lower values of the indicators with low values in errors, i.e., MAE, RRSE, RMSE, and RAE, show better performance (Farooq et al., 2020).

1. $$CC = \frac{m\left(\sum_{i=1}^{m} RS\right) - \left(\sum_{i=1}^{m} R\right)\left(\sum_{i=1}^{a} S\right)}{\sqrt{\left[m\sum_{i=1}^{m} R - \left(\sum_{i=1}^{m} R\right)^2\right]\left[m\sum_{i=1}^{m} R^2 - \left(\sum_{i=1}^{m} R\right)^2\right]}} \quad (11.2)$$

2. $$RMSE = \sqrt{\frac{1}{n}\left(\sum_{i=1}^{n}(S-R)^2\right)} \quad (11.3)$$

3. $$RAE = \sqrt{\frac{\sum_{i=1}^{m}|R-S|}{\sum_{i=1}^{m}(|R-\bar{R}|)}} \quad (11.4)$$

4. $$RRSE = \sqrt{\frac{\sum_{i=1}^{m}(R-S)^2}{\sum_{i=1}^{m}(|R-\bar{R}|)^2}} \quad (11.5)$$

5. $$MAE = \frac{1}{m}\left(\sum_{i=1}^{m}|S-R|\right) \quad (11.6)$$

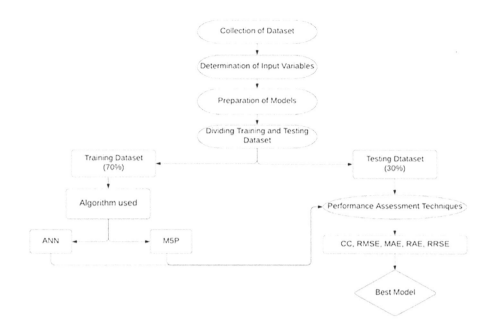

FIGURE 11.2 Flow chart of methodology dataset.

R = Observed values
$\bar{R}$ = Average of observed value
S = Predicted values
m = Number of observations

11.4 RESULT ANALYSIS

11.4.1 Assessment of the ANN-Based Model

In the performance assessment of the ANN-based model, different input variables are measured including CC, RMSE, RAE, RRSE, and MAE and are calculated for the training and testing datasets. The performance evaluation measurement values are shown in Table 11.3 for both stages. Results of Table 11.3 suggest that the statistic quantitative results of ANN with the model number (M6) are performing better than other input combination-based models where CC is 0.8966, MAE is 0.719, RMSE is 0.9376, RAE is 48.69%, and RRSE is 50.18. A scatter graph is drawn (Figure 11.3) for values of actual and predicted ones using the ANN-M6 model for both stages. This figure suggests that the performance of the ANN-M6 based model is suitable for predicting the asphalt concrete mix strength by the Marshall stability test.

11.4.2 Assessment of the M5P Based Model

In the performance assessment of the M5P based model, the combination of different input variables is measured including the CC, RMSE, RAE, RRSE, and MAE,

TABLE 11.3
Performance Evaluation Parameters of an ANN

Model No.	Training					Testing				
	CC	MAE (KN)	RMSE (KN)	RAE (%)	RRSE (%)	CC	MAE (KN)	RMSE (KN)	RAE (%)	RRSE (%)
M1	0.937	0.675	0.8399	41.47	43.52	0.8442	1.1733	1.4279	79.45	76.42
M2	0.9	0.6916	0.908	42.49	47.04	0.7635	1.162	1.5214	78.69	81.42
M3	0.9328	0.6202	0.7879	38.11	40.82	0.8581	1.0642	1.2983	72.06	69.48
M4	0.9321	0.6358	0.8239	39.07	42.69	0.8165	1.125	1.4227	76.18	76.14
M5	0.91	0.5999	0.802	36.86	41.55	0.8507	0.9694	1.1901	65.65	63.69
M6	**0.9272**	**0.5014**	**0.7409**	**30.81**	**38.39**	**0.8966**	**0.719**	**0.9376**	**48.69**	**50.18**
M7	0.8351	0.9775	1.2676	60.06	65.68	0.846	1.2375	1.4476	83.80	77.47
M8	0.8334	0.828	1.0776	50.88	55.83	0.6837	1.2048	1.5358	81.59	82.20
M9	0.8122	0.8473	1.1278	52.06	58.43	0.7149	1.1034	1.3903	74.72	74.40
M10	0.9012	0.68	0.8732	41.78	45.24	0.8694	0.8211	1.026	55.60	54.91

which were calculated for the training and testing datasets. Performance is evaluated and the measured values are listed in Table 11.4 for both stages. Results of Table 11.4 suggest that statistic quantitative results of the M5P-based model with the model number (M3) have better performance in comparison to other input combinations of models having CC (0.8653), MAE (0.75), RMSE (0.9517), RAE (50.79%), and RRSE (50.93%). A scatter graph is drawn (Figure 11.4) for the values of actual vis-a-vis predicted using the M5P-M3 model for both stages. This figure suggests that the performance of the M5P-M3-based model is suitable for predicting the asphalt concrete mix strength by Marshall stability tests.

11.5 COMPARISON BETWEEN BEST DEVELOPED MODELS

The comparison between the ANN and M5P models has been done and the performance of each model was checked for both stages by five statistics evaluation criteria presented in Table 11.5. The scatter graph was plotted (Figure 11.5) for the comparison of actual and predicted Marshall stability values of asphalt concrete mix. In Figure 11.5, the ANN-M6-based model has minimum relative error for the training and testing stages. According to the statistical analysis performance of ANN-M6 and M5P-M3-based models, the ANN-M6 model predicts better results than the M5P-M3 based model. The performance assessment of the ANN-M6 model presents better results in predicting Marshall stability asphalt concrete mix with the highest value of CC (0.8966), followed by M5P-M3 with CC (0.8653); also, the results of RMSE show that ANN-M6 has a minimum error of 0.9376, found to be the best succeeded by M5P-M3 (0.9517). Similarly, RRSE shows minimum error for ANN-M6 (50.18) followed by M5P-M3 (50.93). **The bold number in table 11.3 and 11.4 indicates the best suited model.**

Marshall Stability of Asphalt Concrete Mix 231

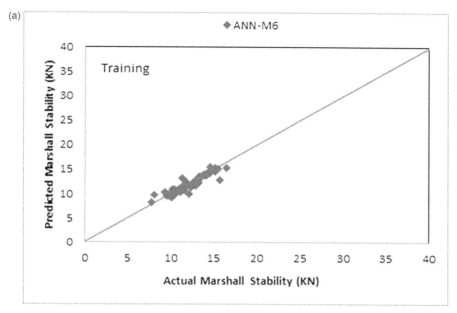

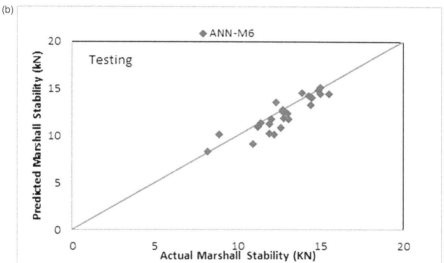

FIGURE 11.3 Scatter graph among actual and predicted Marshall stability by using the ANN model.

TABLE 11.4
Performance Evaluation Parameters of M5P

Model No.	Training					Testing				
	CC	MAE (KN)	RMSE (KN)	RAE (%)	RRSE (%)	CC	MAE (KN)	RMSE (KN)	RAE (%)	RRSE (%)
M1	0.8602	0.7552	1.04	46.40	53.88	0.8583	0.762	0.9444	51.60	50.54
M2	0.8495	0.8172	1.1137	50.21	57.70	0.7888	0.8737	1.1262	59.17	60.27
M3	**0.859**	**0.8044**	**1.0889**	**49.43**	**56.42**	**0.8653**	**0.75**	**0.9517**	**50.79**	**50.93**
M4	0.8635	0.7585	1.0374	46.61	53.75	0.841	0.7914	0.9854	53.59	52.74
M5	0.4238	1.3096	1.7482	80.47	90.58	0.5224	1.2322	1.5711	83.44	84.08
M6	0.8382	0.7792	1.0844	47.88	56.18	0.8008	0.8674	1.0884	58.74	58.25
M7	0.8433	0.8222	1.1308	50.52	58.59	0.7928	0.8664	1.113	58.67	59.56
M8	0.4732	1.302	1.7081	80.01	88.50	0.5041	1.1992	1.5894	81.20	85.06
M9	0.4186	1.3165	1.7528	80.90	90.82	0.4931	1.2363	1.6065	83.72	85.98
M10	0.8286	0.8409	1.1279	51.67	58.44	0.7773	0.9048	1.1565	61.27	61.90

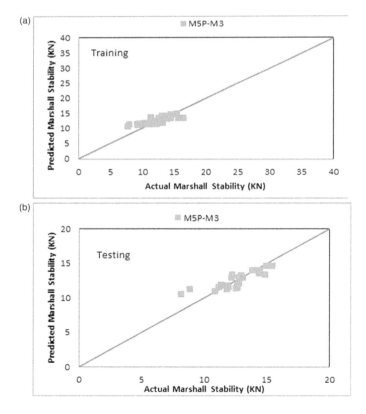

FIGURE 11.4 Scatter graph among actual and predicted Marshall stability by using the M5P model.

TABLE 11.5
Comparison of Models with ANN and M5P

Models	CC	MAE (KN)	RMSE (KN)	RAE (%)	RRSE (%)	Order
ANN-M6	0.8966	0.719	0.9376	48.69	50.18	1
M5P-M3	0.8653	0.75	0.9517	50.79	50.93	2

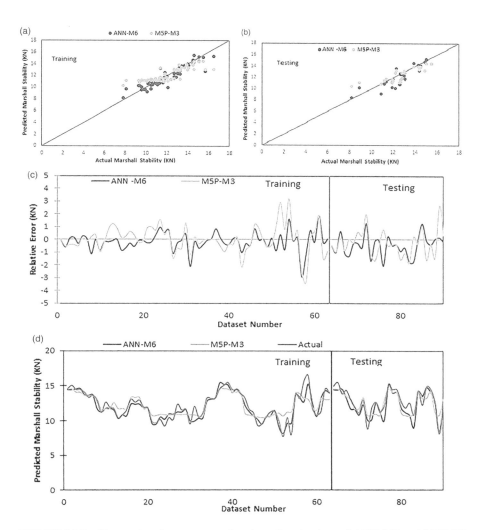

FIGURE 11.5 Comparison between actual and predicted values of ANN-M6 and M5P-M3 models.

11.6 CONCLUSION

In this paper, an investigation has been done to check the potential of the ANN and M5P models to predict the strength of asphalt concrete mix by Marshall stability tests. Four different parameters were applied to evaluate the assessment of models so developed by using soft computing techniques, i.e., ANN and M5P models. The performance evaluation results have been developed by calculating CC, RMSE, MAE, RAE, and RRSE for all ten models. Results of all these techniques are applied for the input parameters, i.e., type of bitumen, BC, voids, VMA, and VFA. It shows that, out of the five prediction models, the performance of the ANN model is more suitable in predicting the Marshall stability of asphalt concrete mix with the value for the CC, MAE, RMSE, RAE, and RRSE being 0.8966, 0.719, 0.9376, 48.69%, and 50.18%, respectively, with the combination of best input parameters, i.e., the type of bitumen, BC, voids, and VMA. Results also show that the ANN-M6 model performance is better due to higher CC followed by the M5P-M3 model. Therefore, it is found that the ANN-M6 model has outperformed in predicting the Marshall stability of asphalt concrete mix.

REFERENCES

Azarhoosh A. and Koohmishi M. Representation of flow number results of hot-mix asphalt using genetic-based model. *Journal of Materials in Civil Engineering*, 2021; 33(2). doi: 10.1061/ (ASCE)MT.1943-5533.0003541.

Behnood A. and Daneshvar D. A machine learning study of the dynamic modulus of asphalt concretes: An application of M5P model tree algorithm. *Construction and Building Materials*, 2020; 262(2020). doi: 10.1016/j.conbuildmat.2020.120544.

Bui D.K., Nguyen T.N. and Chou J.S., Nguyen-Xuan H. and Ngo T.D. A modified firefly algorithm-artificial neural network expert system for predicting compressive and tensile strength of high-performance concrete. *Construction and Building Materials*, 2018; 180, 320–333. doi: 10.1016/j.conbuildmat.2018.05.201.

Dao D.V., Nguyen N.L., Ly H.B., Pham B.T. and Le T.T. Cost-effective approaches based on machine learning to predict dynamic modulus of warm mix asphalt with high reclaimed asphalt pavement. *Materials*, 2020; 13(15), 3272. https://doi.org/10.3390/ma13153272.

Farooq F., Amin N.M., Khan K., Sadiq R.M., Javed F.M., Aslam F. and Alyousef R. A comparative study of random forest and genetic engineering programming for the prediction of compressive strength of high strength concrete (HSC). *Applied Sciences*, 2020; (10), 2–10. doi: 10.3390/app10207330.

Hassan Z, Amir A., Ali M. and Mahdi H. Using the GMDH and ANFIS methods for predicting the crack resistance of fibre reinforced high RAP asphalt mixtures. *Road Materials and Pavement Design*, 2020; 19(1), 1468–0629. doi: 10.1080/14680629.2020.1748693.

Majidifard H., Jahangiri B., Rath P., Contreras L.U., Buttlar W.G. and Alavi A.H. Developing a prediction model for rutting depth of asphalt mixtures using gene expression programming. *Construction and Building Materials*, 2021; 267(1). https://doi.org/10.1016/j.conbuildmat.2020.120543.

Mohammed A., Rafiq S., Sihag P., Kurda R., Mahmood W., Ghafor K and Sarwar W. ANN. M5P-tree and nonlinear regression approaches with statistical evaluations to predict the compressive strength of cement-based mortar modified with fly ash. *Journal of Materials Research and Technology*, 2020; 9(6), 12416–12427. https://doi.org/10.1016/j.jmrt.2020.08.083.

Morova N., Eriskin E., Terzi S., Karahancer S., Serin S., Saltan M. and Usta P. Modelling Marshall Stability of fiber reinforced asphalt mixtures with ANFIS. *Proceedings -2017 IEEE International Conference on Inovations in Intelligent Systems and Applications*, 2017; 174–179. DOI: 10.1109/INISTA.20.8001152.

Nguyen M.D., Pham B.T., Ho L.S., Ly H.-B., Le T.-T., Qi C., Le V.M., Le L.M., Prakash I., Son L.H. and Bui D.T. Soft-computing techniques for prediction of soils consolidation coefficient. *CATENA*, 2020; 195, 104802. https://doi.org/10. 1016/j.catena.2020.104802.

Nhu V.H., Shirzadi A., Shahabi H., Singh S.K., Al-Ansari N., Clague J.J., Jaafari A., Chen W., Miraki S., Dou J., Luu C., Gorski K., Thai Pham B., Nguyen H.D. and Ahmad B.B. Shallow landslide susceptibility mapping: A comparison between logistic model tree, logistic regression, naïve bayes tree, artificial neural network, and support vector machine algorithms. *International Journal of Environmental Research and Public Health*, 2020; 17(8), 2749. https://doi.org/10.3390/ijerph17082749.

Nicola B., Manthos E. and Pasetto M. Analysis of the mechanical behaviour of asphalt concretes using artificial neural networks. *Advances in Civil Engineering*, 2018, 1687–8086. https://doi.org/10.1155/2018/1650945.

Nouri M., Sihag P., Salmasi, F. and Kisi O. Energy loss in skimming flow over cascade spillways: Comparison of artificial intelligence-based and regression methods. *Applied Sciences*, 2020; 10(2020). doi: 10.3390/app10196903.

Ozgan E. Artificial neural network based modelling of the Marshall Stability of asphalt concrete. *Expert Systems with Applications*, 2011; (38), 6025–6030. https://doi.org/10.1016/j.eswa.2010.11.018.

Qadir A., Gazder U. and Choudhary K.U.N. Artificial neural network models for performance design of asphalt pavements reinforced with geosynthetics. *Transportation Research Record*, 2020; 2674(8), 319–326. doi: 10.1177/0361198120924387.

Reddy T.C.S. Predicting the strength properties of slurry infiltrated fibrous concrete using artificial neural network. *Frontiers of Structural and Civil Engineering*, 2018; 12, 490–503. https://doi.org/10.1007/s11709-017-0445-3.

Saffarzadeh M. and Heidaripanah A. Effect of asphalt content on the marshall stability of asphalt concrete using artificial neural networks. *Scientia Iranica*, 2009; 16(1), 98–105.

Saif M.A., El-Bisy M.S. and Alawi M.H. Application of soft computing techniques to predict the stability of asphaltic concrete mixes. *Proceedings of the Third International Conference on Soft Computing Technology in Civil, Structural and Environmental Engineering*, 2013. doi: 10.4203/ccp.103.33.

Serin S., Morova N., Sargin S., Terzi, S. and Saltan, M. The fuzzy logic model for the prediction of Marshall stability of lightweight asphalt concretes fabricated using expanded clay aggregate. *Journal of Natural and Applied Science*, 2013; 17(1), 163–172.

Sihag P., Kumar M. and Singh B. Assessment of infiltration models developed using soft computing techniques. *Geology, Ecology, and Landscapes*, 2020a; 10(20), 2474–9508. https://doi.org/10.3390/app10207330.

Sihag P., Singh B., Sepah-Vand A. and Mehdipour V. Modeling the infiltration process with soft computing techniques. *ISH Journal of Hydraulic Engineering*, 2018; 26(2), 138–152. doi: 10.1080/09715010.2018.1464408.

Sihag P., Tiwari N.K. and Ranjan S. Support vector regression-based modeling of cumulative infiltration of sandy soil. *ISH Journal of Hydraulic Engineering*, 2020b; 26(1), 44–50. doi: 10.1080/09715010.2018.1439776.

Sohel I.S., Singh S., Ransinchung G.D. and Ravindranath S.S. Effect of property deterioration in SBS modified binders during storage on the performance of asphalt mix. *Construction and Building Materials*, 2020. https://doi.org/10.1016/j.conbuildmat.2020.121644

Thakur M.S., Pandhiani S.M., Kashyap V., Upadhya A. and Sihag P. Predicting bond strength of FRP bars in concrete using soft computing techniques. *Arabian Journal of Science and Engineering*, 2021. https://doi.org/10.1007/s13369-020-05314-8.

Vyas V., Singh A.P. and Srivastava A. Prediction of asphalt pavement condition using FWD deflection basin parameters and artificial neural networks. *Road Materials and Pavement Design*, 2020. doi: 10.1080/14680629.2020.1797855.

Ziari H. and Moniri A. Laboratory evaluation of the effect of synthetic polyolefin-glass fibers on performance properties of hot mix asphalt. *Construction and Building Materials*, 2019; 213(2019), 459–468. https://doi.org/10.1016/j.conbuildmat.2019.04.084

Index

Note: *Italic* page numbers refer to figures.

adsorption isotherms 133
ANOVA 183
artificial intelligence (AI) 73
artificial neural network (ANN) 126, 130–*131*, 209, 214, 225, 229
Atomsk 54

bandgap 29; calculation 26, 131
Bragg's law 15

Cleanroom 10
chemical vapor transport (CVT) 4
coefficient of friction 177
composite 72; fabrication 75; flaw detection 128; material constant 134
computational intelligence 125
concrete 194
convolutional neural network 126
critical temperature 127; prediction 127
Czochralski technique 3

deep learning 163
density functional theory (DFT) 20; applications 26
deposition rate 8
dielectric constants 27, 30

effective mass 31
electron-beam lithography (EBL) 11

feature mapping 161; scale-invariant feature transform (SIFT) algorithm 161
fitting 24
functional exchange-correlation 21
fuzzy logic 130, *132*

generalized gradient approximation (GGA) 21; meta-GGA 22

hardness test 83

impact test 83

LAMMPS 55
lattice constant 132; prediction 132
lithography 9
linear regression 185
local density approximation (LDA) 21

M5P tree 196, *200–201*, 226, 229
machine learning 62, 86, 185, 195; image classification 163; material characterisation 161, 188
magnesium silicide 105; analysis 114; bandgap 105; characterisation 110; properties 117; structure 105; synthesis 107; thermodynamic behaviour 106
materials 2; characterization 13; selection 74; synthesis 2; smart laser 129
microscopy 140; atomic force 151; field ion 149; optical 14
molecular dynamics (MD) 50; algorithm 51; force fields 51; parameters 52; property prediction 57

nanofabrication 9

open visualization tool (OVITO) 61

pharmaceutical tableting processes 128, *129*
photolithography 10
physical vapor deposition (PVD) 5
polyamide 6 (PA6) 171; applications 173; features 172; properties 175
polyamides (PA) 171
pull rate 4
pulsed laser deposition (PLD) 8

random forest (RF) 196, *200–201*, 209, 216
random tree 196
range separation 24
rate of wear 179
renewable energy 102
response surface methodology (RSM) 86
roundness test 80

scanning electron microscope (SEM) 13, 143
scanning tunneling microscope (STM) 16, 148
SiO_2 2
soft computing techniques 209
spark plasma 107
spectroscopy 153; energy dispersive (EDS) 159; FTIR 154; optical emission (OES) 156; ultraviolet (UVS) 153
sputtering 7
supervised learning *126*

237

support vector machine (SVM) 195;
    gaussian process regression
    (GPR) 195; multilayer perceptron
    (MLP) 195
supplied test data method 197
surface roughness test 79

Taguchi method 188

thermoelectrics 103; generator (TEG) 103
transmission electron microscope (TEM) 15, 146
turning 76

vapor phase 5
visual molecular dynamics (VMD) 62

X-ray diffraction (XRD) 159

Milton Keynes UK
Ingram Content Group UK Ltd.
UKHW021928181023
430886UK00002B/5